AMATEUR RADIO EQUIPMENT FUNDAMENTALS

AMATEUR RADIO EQUIPMENT FUNDAMENTALS

ALBERT D. HELFRICK

PRENTICE-HALL, INC. Englewood Cliffs, New Jersey 07632

Library of Congress Cataloging in Publication Data

Helfrick, Albert D.
Amateur radio equipment fundamentals.

Includes index.
1. Radio — Apparatus and supplies — Design and construction — Amateurs' manuals. 2. Amateur radio stations. I. Title.
TK9956.H392 621.3841'3 81-1879
ISBN 0-13-023655-1 AACR2

Editorial/production supervision
and interior design by Lori Opre
Manufacturing buyer: Gordon Osbourne
Cover design by Judith A. Matz

Printed in the United States of America

10 9 8 7 6 5 4 3 2 1

Prentice-Hall International, Inc., *London*
Prentice-Hall of Australia Pty. Limited, *Sydney*
Prentice-Hall of Canada, Ltd., *Toronto*
Prentice-Hall of India Private Limited, *New Delhi*
Prentice-Hall of Japan, Inc., *Tokyo*
Prentice-Hall of Southeast Asia Pte. Ltd., *Singapore*
Whitehall Books Limited, *Wellington, New Zealand*

To my children, *Karl* and *Rachel*

CONTENTS

PREFACE

The excitement of amateur radio to many, is the ability to communicate far and wide while giving little thought to what makes that possible. To these amateurs, the circuits and systems behind the front panel are as foreign as a foreign country. All amateurs have the basic knowledge of electronics that was required for their license. The purpose of this book is to take the moderately experienced amateur, that is, one that has advanced beyond the novice stage, through the intricacies of amateur communications equipment. This book covers all phases of communications equipment from simple receivers and transmitters, through sophisticated systems such as slow-scan television and ratioteletype.

The amateur radio operator has the unique advantage of being able to build his own equipment and to use this equipment on the air. No other radio service allows the use of home-constructed equipment. Two chapters contain construction projects for gaining hands-on experience. The projects are of modern design and embody many of the techniques outlined in this book.

The first two chapters cover high-frequency transmitters and receivers, long the mainstay of amateur equipment.

Chapter 3 covers VHF and UHF transceivers. Practically every amateur owns a two-meter transceiver and will find this chapter enlightening. Special consideration is given to synthesizers, a neglected subject in amateur publications.

Specialized forms of communication are covered in Chapter 4. Some very technologically advanced systems are in use by amateurs and some insight is given to the operation of these systems in this chapter.

Even though many amateurs construct their own equipment, most buy some commercial equipment and Chapter 5 tells how to select a transmitter, receiver, or transceiver.

For those on a tight budget, Chapter 6 describes methods of getting more mileage from older, used equipment. The methods described make older equipment look and perform better.

Chapter 7 discusses methods used for the home construction of amateur equipment. The methods described can be used to produce professional-looking equipment in the home workshop.

Chapters 8 and 9 present construction details for transmitters, receivers, and station accessories that can be constructed in the home shop using the techniques described in earlier chapters. The projects described consist of a complete station with choice of receivers and transmitters from a few watts to more than 200-W input using a linear amplifier.

Boonton, New Jersey ALBERT D. HELFRICK

AMATEUR RADIO EQUIPMENT FUNDAMENTALS

1 HIGH-FREQUENCY RECEIVER FUNDAMENTALS

1.1 AMATEUR RADIO STATION REQUIREMENTS

The amateur radio station, historically, consisted of a transmitter, a receiver, and an antenna as a minimum requirement. Recently, this has become a transceiver and an antenna. Add-on accessories such as power amplifiers for increasing the station power, an antenna tuner for improving the match between the antenna and the transmitter, and other accessories, such as keyers, keyboards, speech-processing equipment, and the like, have always been a part of the amateur station. Receiver circuits and systems will be considered separately from transmitter circuits and systems even though the transmitter and receiver are part of a transceiver. Transceivers and the problems associated with the sharing of circuits are discussed at the end of Chapter 2.

1.2 COMMUNICATIONS RECEIVER REQUIREMENTS

Throughout the history of radio, many receiving techniques have been used, but only three have emerged as practical systems. The *superheterodyne* has excelled in both performance and popularity. The *superregenerative receiver*, although simple, suffers some serious electrical drawbacks. Despite this, it saw widespread use during the early days of radio and right up to World War II. Today's crowded amateur bands make the superregenerative almost useless for serious communications. The third technique, the *direct-conversion receiver*, often called the *homodyne*, is actually a special form of superheterodyne. This receiver, not really a high-performance design, has found a place in the amateur radio station because of its simplicity.

All communications receivers must perform three basic functions. The most important is *amplification* of a radio signal. The radio-frequency power as received by an antenna is extremely small, often below 1 picowatt (pW; 10^{-12} watt). To obtain a comfortable listening level from a loudspeaker, the power must be about 1 W. This means that the overall gain of the communications receiver from antenna to loudspeaker must be 10^{12} or about 120 decibels (dB), which is typical of modern communications receivers.

A receiver must be *selective*, able to accept and amplify one radio signal and reject others. Signals entering an antenna feedline number in the hundreds or thousands at many different frequencies and power levels. A receiver must amplify and process only one signal while rejecting all others, even though the desired signal may not be a strong signal. All signal processing of the desired signal by the receiver must be done correctly without interference from one or all of the undesired signals.

Radio communication requires *information* to be impressed on a radio-frequency signal, called the *carrier*, transmitted through the atmosphere to the receiving end, where the information is removed. Thus, a requirement of a communications receiver is to remove this information without error or distortion.

1.2.1 Single-Conversion Superheterodyne

The superregenerative, superheterodyne, and direct-conversion receivers all accomplish each of the three basic requirements of a communications receiver. Because the methods by which each receiver accomplishes the basic functions are different, each receiver has strong and weak characteristics. The most popular and versatile receiver type is the superheterodyne, which was developed during the 1920s and quickly found use in broadcast and communications receivers. Later, superhets, as they are often called, appeared in television receivers, FM equipment, radars, and aviation electronics. Practically every receiver application ever invented made use of the superheterodyne, including the sophisticated communications receivers available today.

Why should the superheterodyne become the most successful communications receiver design and become the mainstay of receivers? It is because the gain of the receiver is accomplished at a low frequency, where the gain per device is high and selectivity is easy to obtain. Previous to the invention of the superheterodyne, one of the more successful receiver designs was the *tuned radio-frequency* (TRF) *receiver.* This receiver was a high-gain radio-frequency (RF) amplifier followed by a detector. Usually, several stages of tuned amplifiers were used before the detector. For low frequencies, such as the broadcast band, a reasonable amount of gain could be obtained without danger of feedback. However, at higher frequencies, the amount of gain per vacuum tube was less and the danger of feedback was always present because even the slightest radiation of a signal from the later stages of the RF amplifier would cause oscillations. Furthermore, obtaining sufficient selectivity from the high-frequency amplifier was impossible.

Figure 1–1 shows a block diagram of a simple superhet communications receiver. The RF amplifier amplifies the signal from the antenna by about 10 to 15 dB and selectivity is included to remove undesired signals called *images.* The signal from the RF amplifier feeds the mixer, where it is hetrodyned with the local oscillator and changed to the *intermediate frequency* (IF). In addition to supplying this frequency conversion, the mixer can supply some of the required signal gain, typically from 10 to 15 dB. Since the total gain requirement is about 120 dB, if the RF amplifier and mixer together provide about 20 to 30 dB of gain, 90 to 100 dB of gain must be obtained elsewhere in the receiver. The way this gain is distributed among the circuits of the communications receiver will greatly affect the performance characteristics of the receiver, and failure to observe good practices will result in poor dynamic range, spurious responses, and poor noise performance. The IF amplifier supplies most of the receiver gain, but even here there are practical limits on the amount of gain that can be reliably obtained. Providing large gain in the RF amplifier, or mixer, is undesirable for several reasons, and there has been a trend in modern communications receiver designs to remove the RF amplifier completely, which is equivalent to providing 0 dB of gain. Most gain-distribution plans call for about 60 dB of gain from the IF amplifier with smaller amounts, about 15 dB each, to be obtained from the RF amplifier and mixer. The remaining gain is from the product detector and the audio amplifier.

Every mixer converts its input frequency to two frequencies simultaneously :

sum frequency: $F_{\mathrm{LO}} + F_{\mathrm{RF}}$

difference frequency: $F_{\mathrm{LO}} - F_{\mathrm{RF}}$

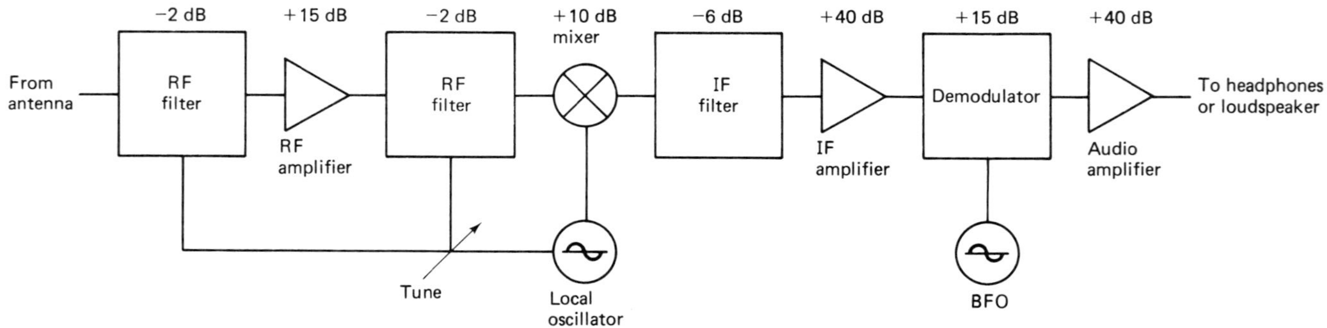

FIGURE 1-1. Simple single-conversion superheterodyne receiver. The gains and losses shown are typical and do not represent the only combination possible.

where F_{LO} is the frequency of the local oscillator and F_{RF} is the frequency of the incoming signal from the RF amplifier. For example, if the incoming frequency is in the 80-meter (m) amateur band, at 3.6 megahertz (MHz), and the local oscillator is operating at 4.055 MHz, the input signal is translated to both 455 kilohertz (kHz) and 7.655 MHz. Since only one of these frequencies can be the IF, a filter is placed at the output of the mixer to pass the 455-kHz signal and trap the 7.655-MHz signal. The IF is chosen very carefully so that the oscillators and the harmonics of oscillators do not combine within the receiver and produce signals that will interfere with the reception of incoming signals. A low-frequency IF is desirable because it is possible to design high-performance filters with standard components and obtain higher gain with fewer active devices.

Since mixers convert an input signal to two IFs, two different input frequencies may be converted to the same IF. One of these signals is called the *desired signal* and the other is called the *image.* For example, using the 80-m receiver as before, if the local oscillator is 4.055 MHz, both 3.6 and 4.51 MHz may be converted to the 455-kHz IF. In this case, the desired frequency is 3.6 MHz and the image is 4.51 MHz. The image must be removed before being fed to the mixer. Because the image is removed from the desired operating frequency by twice the IF, which in the case of the 455-kHz example is always 910 kHz above or below the desired frequency, the lower the IF, the more difficult will be the task of filtering the image. A compromise must be made between image elimination and IF selectivity when choosing an IF.

The frequency of the RF amplifier and the local oscillator differ by the IF regardless of the input frequency. Most receivers have a single tuning control that adjusts the RF amplifier and the local oscillator so that they are always separated by exactly the IF. Special capacitors are used that can be carefully adjusted to have perfect alignment, which is called *tracking.*

The IF selectivity is always placed between the mixer and the IF amplifier to remove all the interfering signals. It is always good practice to remove as many undesired signals as possible before an IF amplifier so that there will be less interference between the desired and the undesired signals. This is especially true in amplifiers where there is high gain. In addition to providing the bulk of the receiver gain, the IF amplifier provides automatic gain control, noise elimination, and the S-meter. Bipolar transistors, field-effect transistors (FETs), and integrated circuits all make good IF amplifiers, and if the frequency is low, one integrated circuit is all that is required for a complete IF amplifier.

The *demodulator,* which immediately follows the IF amplifier, satisfies the last of the three requirements of the communications receiver. The demodulator removes the information from the received signal, and its circuit depends on the type of modulation being received. Many communications receivers are capable of receiving several different types of modulation and hence have several different demodulators. For single-sideband (SSB) and continuous-wave (CW) work, the demodulator must have a *beat-frequency*, or carrier, *oscillator* (BFO). A product detector heterodynes the IF signal with the beat-frequency oscillator and produces a demodulated SSB signal or an audible beat note for CW copy. The product detector, being basically a mixer, can provide some of the required gain as well as provide low-distortion demodulation.

Because the signal level at the output of the demodulator is not sufficient for driving headphones or a loudspeaker, an audio amplifier increases the level to 1 W or so with a maximum distortion of 1 to 5 percent. Some additional selectivity from an audio filter may be included in the audio amplifier, but such filtering is never an adequate substitute for IF selectivity.

1.2.2 More Complex Variations of the Superheterodyne

An expanded block diagram of the simple superheterodyne used as an example is shown in Figure 1–2, with the controls and an *automatic gain control* (AGC) system included.

AGC automatically controls the gain of the receiver to accommodate changes in received signal strength. This system keeps the audio level at the headphones or loudspeaker at a constant level for signals of different strengths or for variations in one signal due to atmospheric fading. The constant audio level prevents operator fatigue and the need to constantly adjust the gain of the receiver manually.

The RF gain control varies the gain of the RF amplifier and is used during CW and SSB operation. In older receivers, the AGC was disabled when the BFO was operating and the gain of the receiver was controlled manually with the RF gain control for CW and SSB operation. Modern receivers have effective gain-control systems that operate with the BFO, which allows the RF gain control to be set and the audio gain control to be used as a simple volume control. Some operators enjoy using the RF gain control even though their receiver has an effective AGC system.

In a receiver having more than one IF filter, the bandwidth can be set by the selectivity switch. At least three filters are required for a communications receiver used for amplitude modulation, (AM), SSB, and CW. A bandwidth of 2.1 kHz is desirable for SSB, 6 kHz for AM, and 500 Hz for CW. Many high-performance receivers have provision

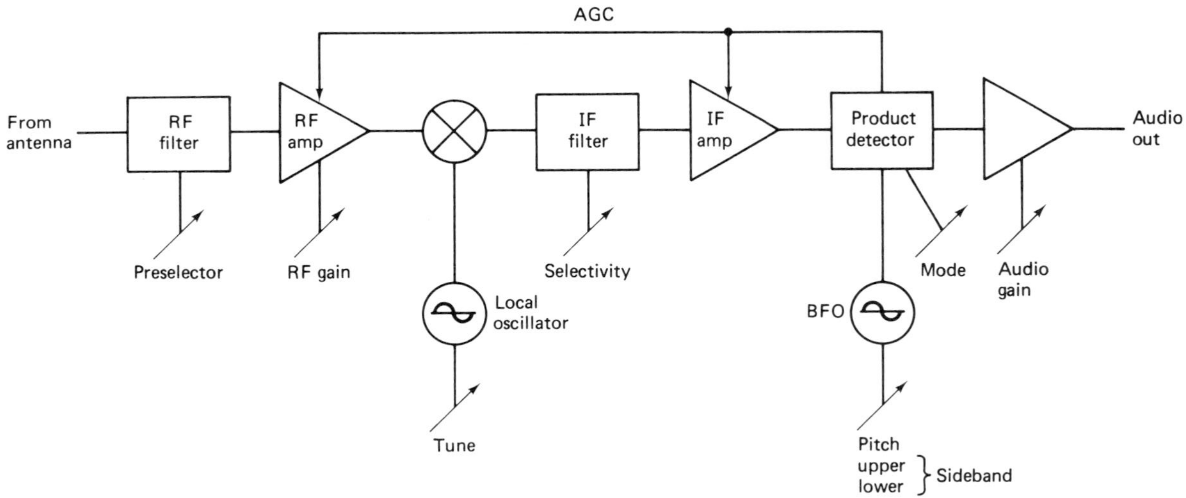

FIGURE 1-2. Simple single-conversion superheterodyne using a separate preselector and showing the front-panel controls and the circuits they affect.

for additional filters, such as 1.8 kHz for crowded-band SSB or 1000 Hz for radioteletype.

Demodulator selection is accomplished by the mode switch. In addition to selecting the demodulator, the mode control can also enable the BFO and select the optimum IF bandwidth.

The pitch control varies the frequency of the BFO, which changes the pitch of the beat note for CW and is used to select either upper or lower sideband for SSB operation. Receivers with crystal-controlled BFOs would not have a pitch control unless a separate variable-frequency BFO is supplied.

Dual-Conversion Superheterodyne. Many variations of the simple superheterodyne have evolved to satisfy specific requirements or to enhance the performance of the receiver. One of the more widely used refinements is dual conversion, shown in Figure 1-3.

Dual conversion is an attempt to reap the benefits of both a high and a low intermediate frequency. The high IF allows good image performance. Since the image is removed from the desired frequency by twice the IF, making the IF as high as possible will move the image as far away from the desired frequency as possible and make the job of filtering the image easier. A second IF is chosen to be much lower in frequency to take advantage of high-Q resonators and narrow-band filters for good selectivity.

The actual received frequency is set by the frequency of both local oscillators, which is the basis for another advantage of the dual-conversion receiver. The second local oscillator is made very stable by keeping the frequency low and restricting the tuning range to several hundred kilohertz. Since the tuning range is narrow, an accurately calibrated dial may be affixed to the second local oscillator. The first local oscillator is then made crystal-controlled for stability. The band-switch selects the frequency of the first local oscillator by switching crystals, and bandspread is accomplished by tuning the second local oscillator. Very stable and accurate receivers have used this technique for many years.

A serious disadvantge is the large number of crystals required for the first local oscillator. Modern designs now use a frequency synthesizer instead of a crystal oscillator, but the synthesizer used for this purpose must have excellent noise characteristics and be carefully shielded to prevent radiation of spurious signals.

Although dual-conversion receivers are among the better-performing receivers today, they are not free of problems. Even though the high IF was chosen for the purpose of image reduction, the second mixer has its own image and, if not properly suppressed, can actually increase the image problem. A second potential hazard is internally generated

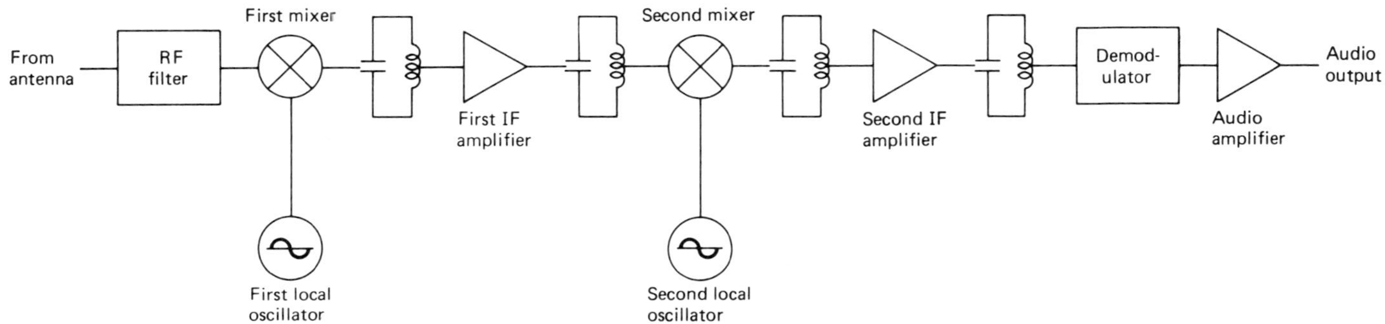

FIGURE 1-3. Dual-conversion superheterodyne without an RF amplifier. The first mixer may be a double-balanced mixer for improved dynamic range.

spurious signals. The dual-conversion receiver has two local oscillators, two IFs, and a BFO, which increases the possibility that some combination of those frequencies will produce a spurious signal. In a well-designed dual-conversion receiver, the IFs are carefully chosen and the receiver is constructed with ample shielding to prevent spurious responses. In spite of all the potential for problems, there are many superheterodyne designs with dual and even triple conversion.

Up-Converting Superheterodyne. A special case of dual-conversion superheterodyne is the up-converting receiver. The term *up-converting* is used because the first IF is very high, often in the very high frequency (VHF) range and is always above the highest frequency covered by the receiver. Recalling that the image is removed by a frequency equal to twice the IF, the image will always be more than twice the highest input frequency. Consider the example shown in Figure 1-4. This receiver has a first IF of 48.05 MHz and covers the input frequency range 0 to 30 MHz. The band of frequencies covered by the image is from 96.1 to 126.1 MHz, which can be very easily eliminated with a low-pass filter with a cutoff slightly above 30 MHz. Only one filter is needed to eliminate the entire image band and pass the input frequency range. This filter need not be tuned with the local oscillator and does not have to track. Since a large band of frequencies appears at the first mixer, great care must be taken in the design of this mixer for low distortion.

Since it is difficult to build a 48-MHz filter narrow enough for CW or SSB operation, the first IF must be converted to a lower frequency such as 5645 kHz to obtain the required selectivity.

In this example, the image from the second conversion is removed

The Drake R7 receiver is a high-performance general coverage receiver used by amateurs as an auxiliary receiver. (Courtesy of R. L. Drake Company.)

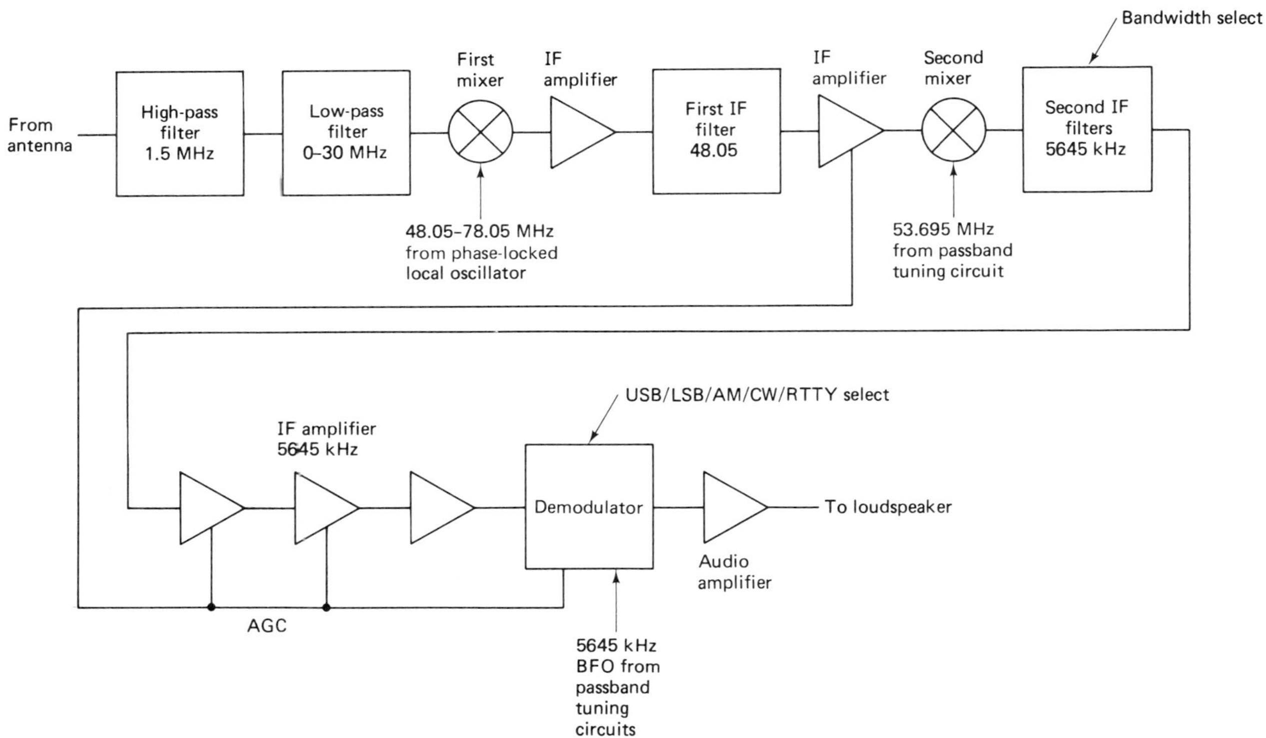

FIGURE 1-4. Typical up-converting receiver using no RF amplifier. This circuit is a simplified block diagram of the receiver section of the R. L. Drake TR7 high-frequency transceiver.

from the first IF by 11.29 MHz and must be removed by the first IF filter. A monolithic crystal filter or *surface wave* (SAW) *filter* is ideal for use as a first IF filter.

The first local oscillator must cover the range 48.05 to 78.05 MHz with an accuracy and stability good enough for SSB tuning. Typically, the tuning accuracy is better than 1 kHz, whereas stabilities of better than 100 Hz for 1 hour are acceptable for SSB. This is not possible with an *LC* (inductor–capacitor) oscillator and must be provided by a frequency synthesizer. A frequency synthesizer suitable for an up-converting receiver is a sophisticated device often using multiple loops and other mixing schemes and is beyond the scope of this book. Some insight into frequency synthesizers is presented in Chapter 3.

1.2.3 Local Oscillator Systems

In spite of the great use of multiple-conversion receivers, there are major advantages of single conversion. Because the single-conversion receiver has only one mixer, one local oscillator, and a carrier oscillator, the number of spurious signals is low and the dynamic range is great. Modern crystal-filter technology has produced excellent lattice filters in the 5 to 9 MHz range, an IF high enough to provide good image performance. The key to a good performing single-conversion communications receiver is generating a stable, accurately calibrated local oscillator without producing spurious signals within the receiver. Heterodyning a variable oscillator with a crystal oscillator as shown in Figure 1–5 is one method for achieving the required local oscillator. The mixer must be shielded and the filters must have good stop-band characteristics; otherwise, undesired signals will escape and cause spurious responses in the receiver. The crystal oscillator and the variable oscillator frequencies must be chosen very carefully so that none of the harmonics fall on the IF or an image frequency. Digital frequency readout may be applied to the variable oscillator and the display manipulated digitally to provide the proper readout.

A more elegant method being used more often in modern receivers is a *phase-locked loop.* Phase-locked-loop methods often require only one crystal and can easily be made free of spurious signals. A voltage-controlled oscillator drives the first mixer as well as a mixer for the phase-locked loop as shown in Figure 1–6. A crystal oscillator is heterodyned with the voltage controlled oscillator (VCO) and the difference frequency is compared to a variable-frequency oscillator in a phase detector. The error voltage from the phase detector is fed back to the VCO to close the loop. It is much easier to reduce the levels of spurious signals in this circuit than in the heterodyning circuit described earlier. The only input to the VCO is a low-frequency line for voltage

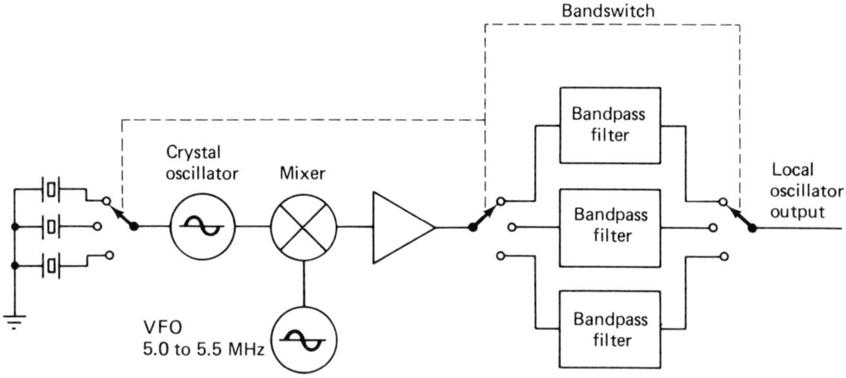

RCV frequency	Local oscillator frequency	Crystal frequency
14.0 to 14.5	23.0 to 23.5	18.0
21.0 to 21.5	30.0 to 30.5	25.0
28.0 to 28.5	37.0 to 37.5	32.0

FIGURE 1-5. Mixing type of local oscillator. Shielding must be used to assure spurious-free operation.

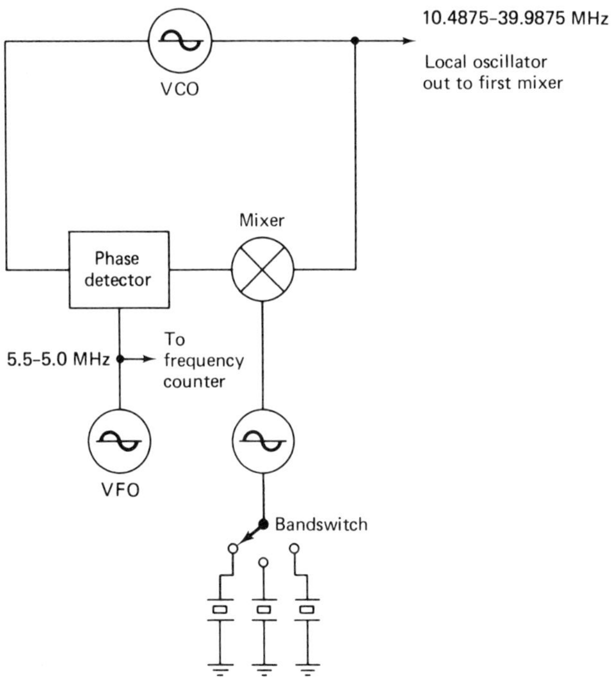

FIGURE 1-6. Phase-locked-loop local oscillator for a SSB transceiver. Notice that only the crystal is changed during a bandswitch.

tuning. It is not difficult to shield the components of the phase-locked loop to reduce spurious signals to safe levels. As with the heterodyning type of local oscillator, a digital readout can be used in conjunction with the variable-frequency oscillator.

1.2.4 Direct-Conversion Receiver

The homodyne or direct-conversion type of receiver is a single-conversion superheterodyne with a very low (audio frequency) IF. It contains only a BFO for frequency conversion and an audio amplifier for gain, as shown in Figure 1-7.

As with all superheterodynes, the input RF signals are filtered to remove some of the undesired signals. Images in the usual sense are not present, so that the input RF filter does not have to be unusually narrow. The incoming signal is beat with the local oscillator with what is essentially a product detector, which may be an active type using transistors, an integrated circuit that will provide some gain, or the diode type with good distortion characteristics but which introduces a loss. The demodulated signal from the product detector is very small, especially if the diode mixer is used, and all the gain is obtained at audio frequencies. Producing gain at audio frequencies has two significant advantages: (1) transistors and integrated circuits provide higher gains at lower frequencies, and (2) active or *LC* filters can give the required selectivity with standard components.

Although a direct-conversion receiver is useful for CW, SSB, radioteletype (RTTY), and slow-scan television, and can be constructed to have a decent noise figure, sensitivity, and dynamic range, it suffers from one serious drawback: signals may be heard on both sides of zero beat. Very early superheterodynes suffered from this problem, which was merely a lack of IF selectivity. Later receivers with improved selectivity were called *single-signal* superheterodynes, to indicate that only one beat note could be heard. These receivers were introduced

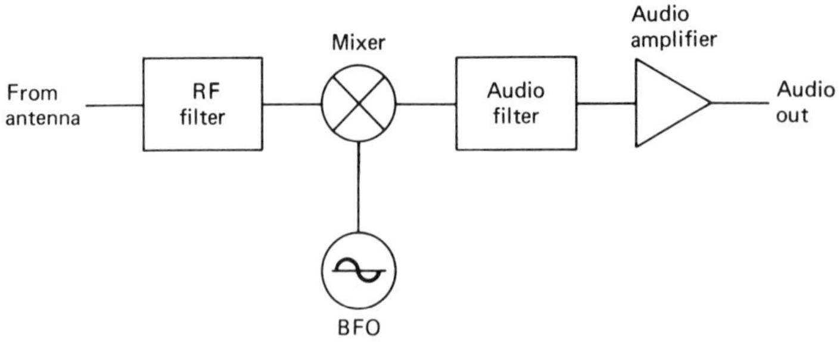

FIGURE 1-7. Direct-conversion receiver. Most of the gain of this receiver is obtained from the audio amplifier.

before the days of single sideband, so this characteristic was an indication of a good CW receiver. Although the improved selectivity was helpful for AM reception, the single-signal concept was applied only to CW.

For CW operation, the direct-conversion receiver is a single-conversion superheterodyne with an IF of about 750 Hz, the frequency of the beat note. Therefore, there exists an image removed by twice the IF, or 1500 Hz, which cannot be removed by a conventional RF filter. Since the image is always present, the CW signal can be heard on both sides of zero beat, or when several signals are present simultaneously, all signals within 1500 Hz of the tuned frequency can be heard. Therefore, the direct-conversion receiver does not meet the requirements for single-signal reception.

When receiving single sideband, the IF is essentially zero, and the receiver will respond to both sidebands. If double sideband or normal AM were the mode of operation, there would be no spurious responses. With a single sideband, the space normally occupied by the unwanted sideband is the adjacent channel and is a source of interference in the direct-conversion receiver.

In spite of this serious flaw, the direct-conversion receiver is attractive because it possesses many good characteristics with relatively simple circuits.

1.2.5 Superregenerative Receiver

The final receiver type to be considered is presented for historical rather than practical reasons. The superregenerative receiver offers good sensitivity with extreme simplicity, but suffers from poor selectivity, a tendency to overload, and can radiate energy from the antenna.

It can be noted from the block diagram in Figure 1-8a that the receiver is extremely simple, even with the addition of the RF amplifier, which is practically a requirement to isolate the antenna from the oscillating detector.

In Figure 1-8b, an FET version of the superregenerative detector is shown. Because of feedback similar to an oscillator circuit, the detector oscillates at the frequency of the signal received. The oscillations cannot last indefinitely, owing to a change in bias voltage because of the oscillations. After a short burst of oscillation, the receiver is said to be quenched and the oscillations desist. In a short period of time, thermally generated noise will cause the oscillations to start again and the process is repeated. Coupling energy into the circuit from the antenna causes the starting point of the oscillations to change as a result of the added energy from the input signals. This change in starting point can be coupled out of the detector circuit with an audio trans-

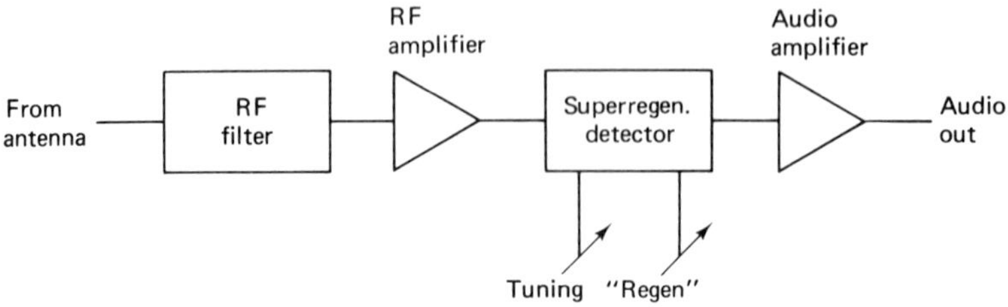

(a)

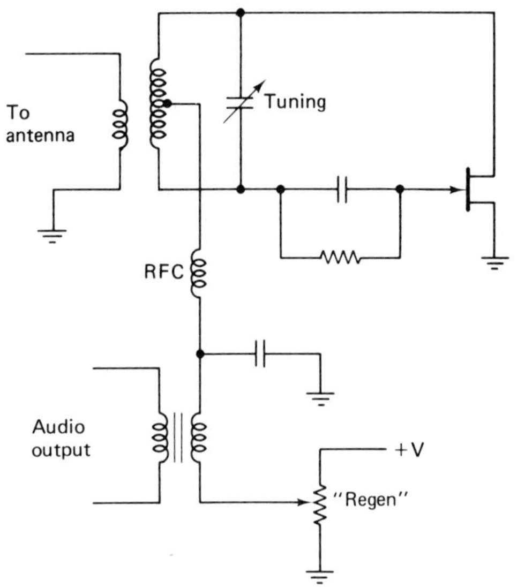

(b)

FIGURE 1-8. (a) Block diagram of a superregenerative receiver, showing its extreme simplicity; (b) schematic of an FET superregenerative detector, the heart of the superregenerative receiver.

former and further amplified. This circuit is effective for demodulating AM, mildly successful for detecting CW or SSB, and of no value for frequency modulation (FM). With the proper transistors and tuned circuits, the superregenerative may be used up into the microwave region.

1.3 RECEIVER CIRCUITS

Although there are great differences between the block diagrams of the various receivers described, they are all constructed with the same basic circuits. The circuits discussed in the next few pages are representative of circuits in use and do not represent all possible variations or indicate that the circuits shown are superior to any other design.

1.3.1 RF Amplifiers

RF amplifiers in use today consist of three types: the common emitter, the common base, and the cascode. The field-effect counterpart for the common emitter is the common source, and for the common base is the common gate. Common-collector and common-drain amplifiers, which can theoretically supply gain, are seldom, if ever, used. High gain and low distortion can be obtained from the common-emitter and common-source amplifiers shown in Figure 1-9 if the proper active device is used. The bipolar transistor shown in Figure 1-9a is biased to operate class A by the bias voltage divider of R1 and R2. Tapping down both the base and antenna connection preserves the loaded Q of the tuned circuit because of the relatively low impedance of both the base and antenna. The collector load is somewhat higher and does not always need to be tapped down.

Gain control may be applied by varying the bias current of the circuit, which effectively changes the gain by varying the operating point of the transistor. Special transistors characterized for gain-control amplifiers have a well-defined gain versus bias current relationship and should be used in gain-controlled RF amplifiers.

Great care must be taken in the construction and layout of this circuit to prevent self-oscillations. The slightest feedback can cause oscillations at the operating frequency and in some cases at very high frequencies far removed from the frequency of the tuned circuits. In many cases, negative RF feedback can be added to the circuit to achieve an unconditionally stable amplifier, a technique called *neutralizing.*

Figure 1-9b shows the equivalent field-effect transistor version or the common-source amplifier. Since the input impedance of the FET is much higher than the bipolar transistor, only the antenna circuit is tapped down on the input circuit. Careful layout is required and in some cases neutralization is required. The addition of gain control is difficult, if not impossible, with junction FETs.

There are special transistors manufactured for use in receiver RF amplifiers and include bipolar, junction FET, and MOSFETs. These transistors are characterized by having a low noise figure while retaining a high signal capability. Using such a transistor to its advantage involves very high bias currents and optimum impedance matching. It is not unusual to have a receiver RF amplifier dissipating as much as 1 W. The dynamic range of such a circuit can be as high as 100 dB.

Some of the features of the dual-gate MOSFET are used in the RF

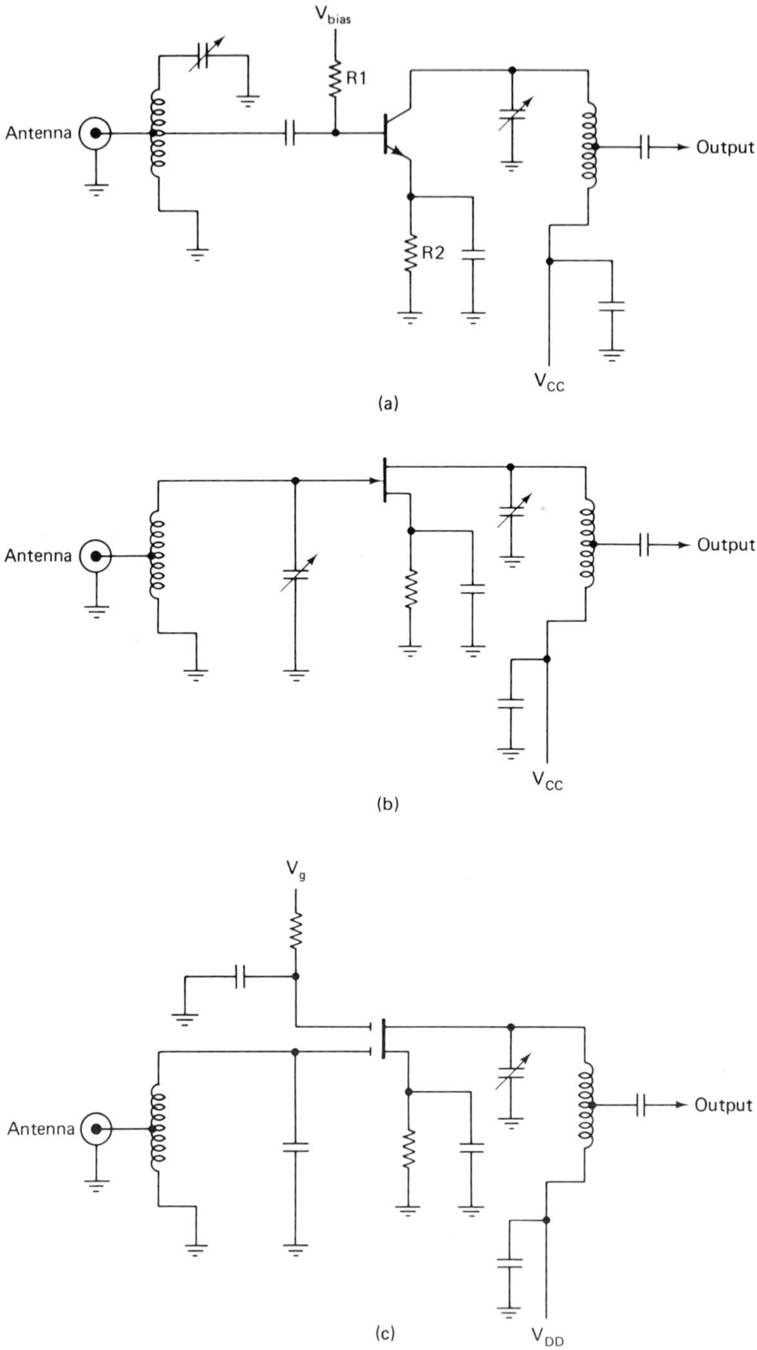

FIGURE 1-9. RF amplifiers using bipolar and field-effect transistors.

amplifier shown in Figure 1-9c. The second gate is an excellent source of gain control, and because of the lower feedback capacitance, neutralization is not usually required. The gain and noise figure are as good as any junction FET or bipolar transistor, but the dynamic range of the dual-gate MOSFET is not as good.

Figure 1-10a shows the FET version of the common-gate RF amplifier. The gain of this circuit is on the order of 10 to 14 dB, which is somewhat less than that which is obtained from the common-source and common-emitter RF amplifiers. The main advantage of the common-gate amplifier is that it almost never requires neutralization, even into the VHF spectrum. Unlike the common-source amplifier, the input impedance is low, usually on the order of 50 to 75 ohms while the output impedance is similar to the common-source configuration.

The bipolar transistor version, the common-base amplifier, has

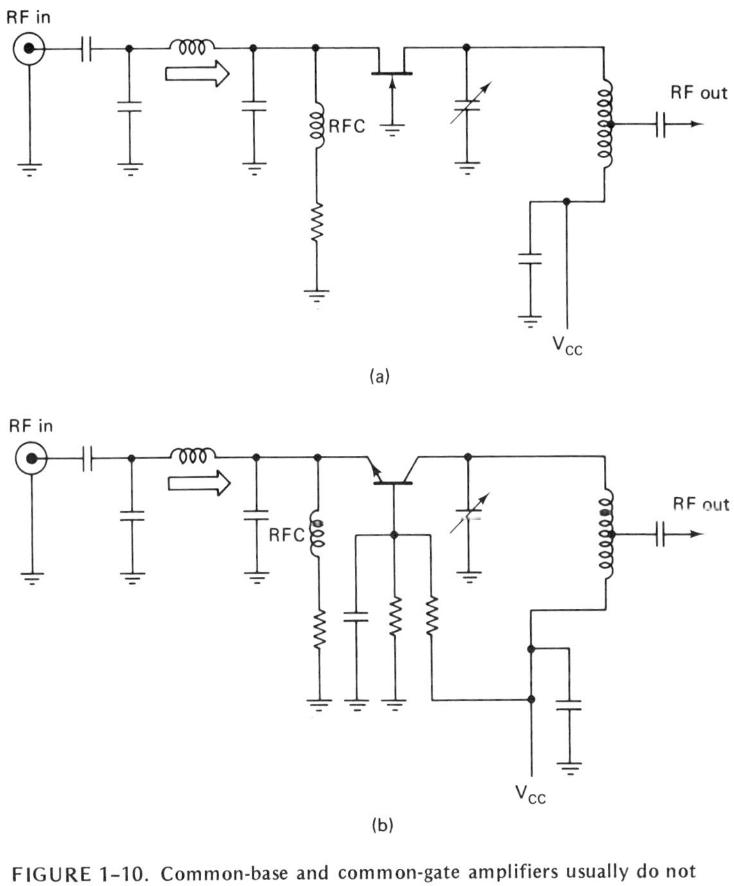

FIGURE 1-10. Common-base and common-gate amplifiers usually do not require neutralizing.

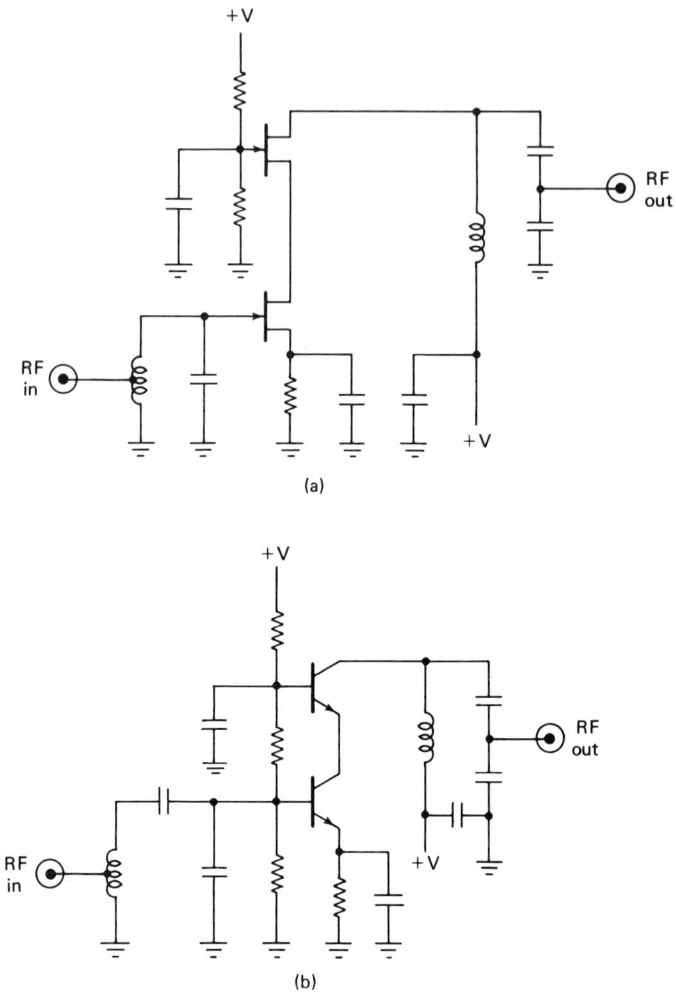

FIGURE 1-11. (a) Field-effect cascode amplifier; (b) the bipolar version of the cascode amplifier.

similar characteristics and also requires no neutralization. Both of these circuits are excellent for home-constructed equipment because almost any reasonable layout produces an unconditionally stable amplifier from very low frequencies to several hundred megahertz.

The major disadvantage of the common-base and common-gate circuit is the difficulty in providing gain control, which must be supplied from an external circuit such as a PIN diode attenuator. Usually, the amplifier is operated without gain control.

The cascode amplifier, as shown in Figure 1-11, is essentially a common-emitter amplifier driving a common-base stage and supplies very high gain without neutralization, because of the low feedback through two devices. Gain control may be applied to the first stage by adjusting the bias voltage. Since the two stages are in series for any direct currents, changes in the first-stage current affect the second stage. The FET version of the cascode amplifier, shown in Figure 1-11b, is similar in performance to the bipolar version except that gain control is difficult to apply. Both FET and bipolar cascode circuits require relatively high supply voltages, since the supply voltage must be divided between two transistors. This can be a significant problem in mobile or portable equipment with limited power supply voltages, such as encountered in a 12-volt automotive application.

1.3.2 Mixers

Mixers fall into two broad categories, active and passive. *Active mixers* use transistors or integrated circuits and usually provide some signal gain; *passive mixers* use diodes and cause a signal loss. Significant advantages in dynamic range and intermodulation distortion of diode mixers cause them to be used often in high-performance communications receivers in spite of the signal loss.

Mixers can be further divided into balanced and unbalanced types. *Balanced mixers* are constructed such that there is an inherent isolation between the local oscillator and the RF or IF ports, or in the case of the doubly balanced mixer, isolation occurs between any two ports. *Unbalanced mixers* have varied amounts of isolation depending on the type of circuit used. In some cases, such as the dual-gate FET mixer, the local oscillator to RF isolation can be as high as 35 dB, but in the FET gate injection mixer the isolation is 0 dB.

In a modern high-frequency communications receiver where there is no RF amplifier, good isolation in the mixer between the local oscillator and the antenna is required to keep the local oscillator signal from being radiated by the antenna. The RF to IF isolation prevents *IF breakthrough*, which occurs if signals in the IF range leak through the mixer and are amplified and detected by the IF amplifier. This is especially troublesome when the IF is in the high-frequency spectrum, such as 10.7 MHz, where signals near the IF are often very strong and require a lot of isolation to prevent interference.

Figure 1-12 shows a double-balanced diode mixer of the type that is most often found in a commercial packaged mixer. The performance of this mixer depends on the electrical balance between all the mixer components. The forward voltage drop of the diodes must be matched as well as the diode capacitance. Hot carrier, or Schottky barrier diodes,

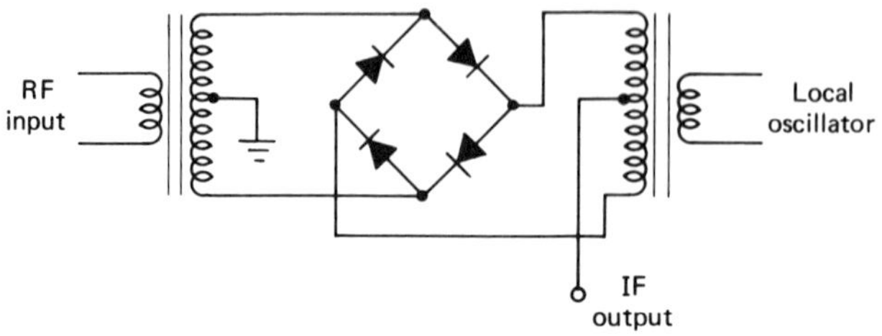

FIGURE 1-12. Typical wide-band double-balanced mixer used in many modern communications receivers.

have inherently better matching, and integrated diode chips have superior performance. The transformers must be absolutely symmetrical and are often bifilar-wound to ensure tight coupling. This type of mixer is capable of good performance over a wide frequency range if correctly constructed.

Several modifications of the basic mixer circuit are used to increase the dynamic range of the mixer. One such modification is shown in Figure 1-13. Here the number of diodes has been increased by a factor of 2. Even larger numbers of diodes may be used for even greater dynamic range with an increased local oscillator power requirement. Preventing interference from the high local oscillator power can be an almost impossible problem, so that the very high level mixer is used only in systems where it is relatively easy to isolate the local oscillator from the antenna.

An active balanced mixer, as shown in Figure 1-14, has all the isolation characteristics of the diode balanced mixer but provides signal gain instead of loss. If a balanced pair of high-quality FETs are used, this circuit can provide superior intermodulation performance and dynamic range, making this mixer one of the best available.

Figure 1-15 shows a double-balanced mixer using an integrated

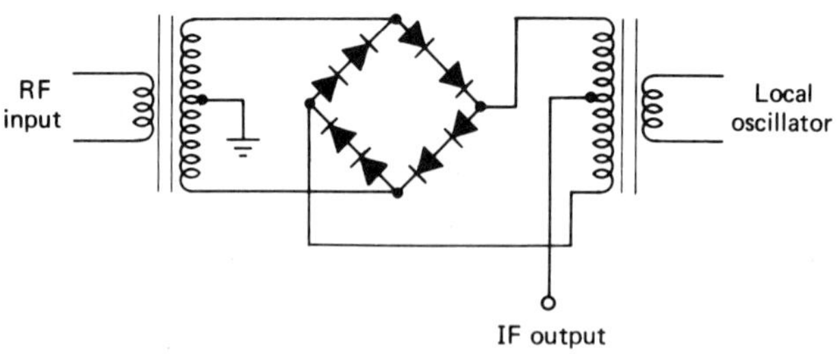

FIGURE 1-13. High-performance double-balanced mixer suitable for receiver use. The number of diodes may be increased to practically any amount with a subsequent increase in local oscillator level.

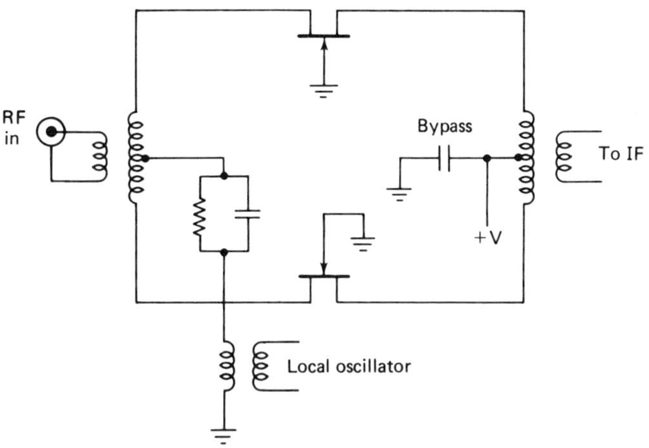

FIGURE 1-14. The active balanced mixer offers excellent isolation, superior dynamic range, and gain.

circuit. Because of the inherent matching of the transistors in an integrated circuit, the balance, and hence the isolation, of this mixer is excellent. Because of the smaller size of the devices used in the integrated circuit, the dynamic range and the intermodulation performance, although acceptable, are inferior to that of Figure 1-14.

Two single-ended or unbalanced-type active mixers are shown in

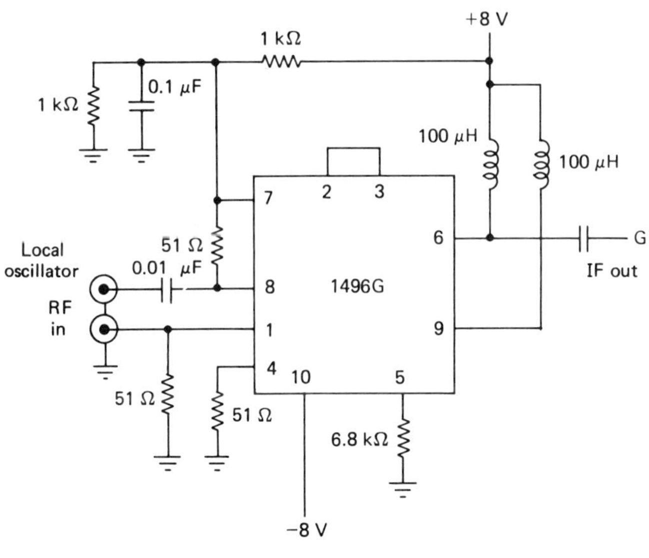

FIGURE 1-15. A balanced mixer that offers good isolation and good gain.

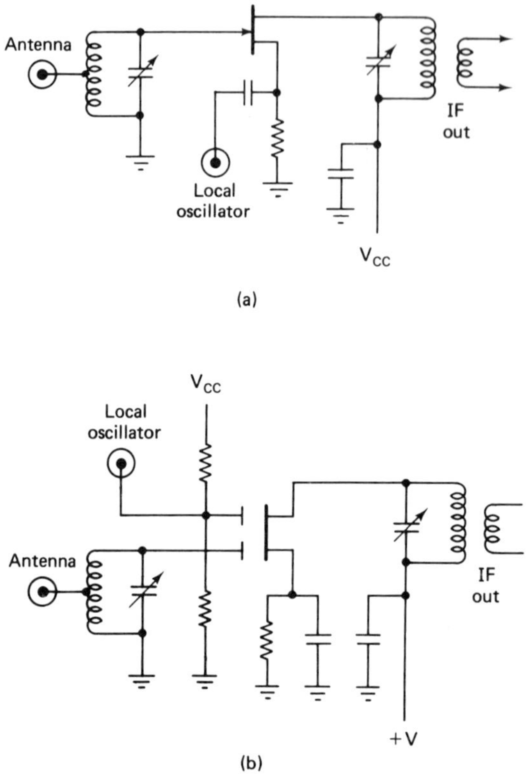

FIGURE 1-16. Two simple inexpensive single-ended mixers for noncritical applications.

Figure 1-16. The dual-gate MOSFET mixer shown in Figure 1-16b is the best single-ended mixer and a very popular circuit. The local oscillator to RF isolation is good but the RF to IF and the local oscillator to IF isolation is poor. It is imperative that a good-quality filter immediately follow the mixer circuit for good spurious-free reception. It is also imperative that the RF filter ahead of the mixer have a high attenuation to the IF to prevent breakthrough. High levels of conversion gain may be obtained from this circuit with inexpensive components, while preserving an acceptable dynamic range and intermodulation performance.

In the junction FET single-ended mixer shown in Figure 1-16a the local oscillator is injected into the source while the RF signal is fed to the gate. Similar results could be obtained if the local oscillator and RF signal were both applied to the gate. This would reduce the local oscillator to RF isolation, which would be of extreme importance if there were no RF amplifier in the receiver. It is important to have a sufficient local oscillator level in this mixer because a low level degrades the noise figure, intermodulation performance, and dynamic range. If properly applied, the junction FET mixer will perform as well as the dual-gate MOSFET.

1.3.3 Filters

IF filters are classified by the type of resonator used, and include *LC* (inductor–capacitor), mechanical, ceramic, and crystal filters.

LC filters are used only at very low IFs, such as those below 100 kHz, in inexpensive receivers, or in very old receivers built before the days of single sideband. The high *Q*'s required for good skirt selectivity required for SSB operation mandates a very low frequency IF and large inductors. Some older receivers used double and even triple conversion with several tuned circuits at the lowest IF. The problems associated with double and even triple conversion are hardly worth the cost saving from the use of *LC* filters. Modern low-loss crystal and mechanical filters have all but eliminated the need for any *LC* filters in the communications receiver. Simple *LC* filters are still used in very inexpensive shortwave receivers suitable only for shortwave broadcast reception.

Ceramic filters are most often used at frequencies below 500 kHz, with 455 kHz being a universal standard. Filters are available from simple single-section filters to 10-section or larger ladder filters suitable for serious communications receivers. Although not inexpensive, the cost of a 10-section ceramic filter can be less than a comparable mechanical filter. Wide-band ceramic filters have been manufactured for frequencies as high as 10.7 MHz for use in FM broadcast receivers, but narrow-band filters for SSB for high IFs are not available. The low loss, good skirt selectivity, and low price make ceramic filters attractive for moderately priced communications receivers.

Crystal filters are the oldest type of high-performance filter. Even early communications receivers had a crystal filter with the ubiquitous phasing control that, if properly manipulated, could provide wondrous amounts of selectivity. Modern crystal filters use several crystals and are computer-designed to achieve excellent selectivity. Modern crystal filters are available with center frequencies up to about 20 MHz using separate crystals and up to 150 MHz using monolithic filters. Bandwidths can vary from a low of less than 100 Hz to about 6 kHz for AM operation. Because of the extremely high *Q*'s involved, the loss of a crystal filter is generally the lowest of all filter types.

1.3.4 Demodulators

The simplest demodulator is the *half-wave rectifier* used for AM detection, shown in Figure 1–17. RF alternating currents are rectified through the diode, D1, and the ripple is filtered with C1. Although

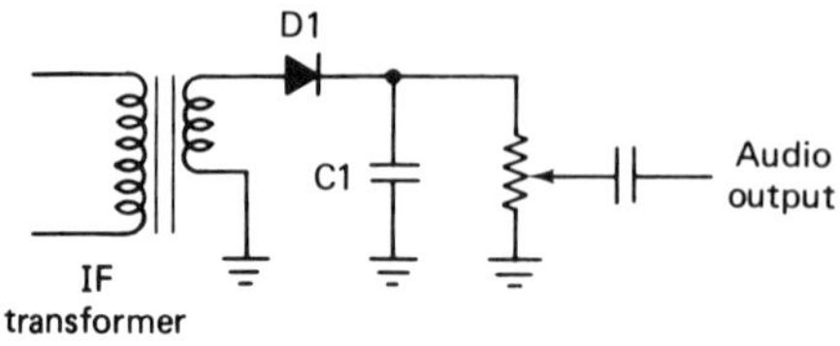

FIGURE 1-17. This simple diode detector is suitable only for AM signals.

good linearity and low distortion are possible, this circuit provides no gain. Extreme simplicity has made this circuit popular for inexpensive receivers. By proper application of a BFO signal, the simple diode detector can be used for CW and SSB operation, but suffers from distortion and poor conversion gain.

A product detector using an active balanced mixer integrated circuit (IC) is shown in Figure 1-18. Because the product detector is nothing more than a mixer with an audio-frequency output, there is a great similarity to the mixer circuit shown in Figure 1-15. Several requirements for a mixer are relaxed for a product detector. Intermodulation is seldom a problem because most of the interfering signals have been removed by the IF filter. Dynamic range is of less importance because the AGC system has stabilized the signal levels, and isolation between various ports of the mixer is not critical because there is no chance of

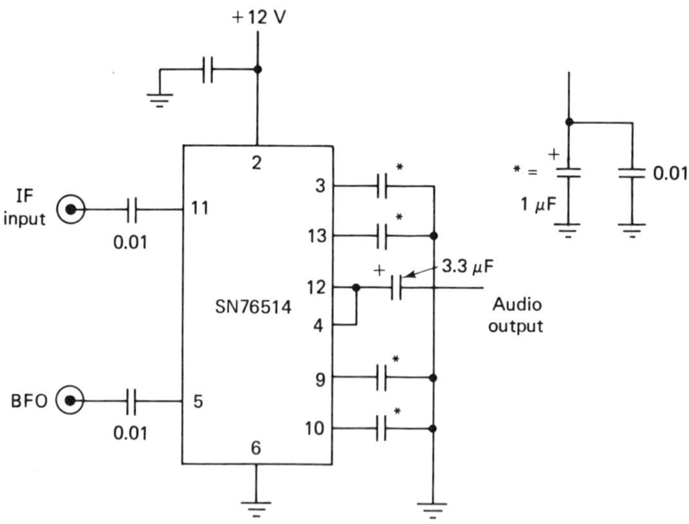

FIGURE 1-18. The integrated-circuit double-balanced mixer contains all the components required for the product detector except the bypass capacitors.

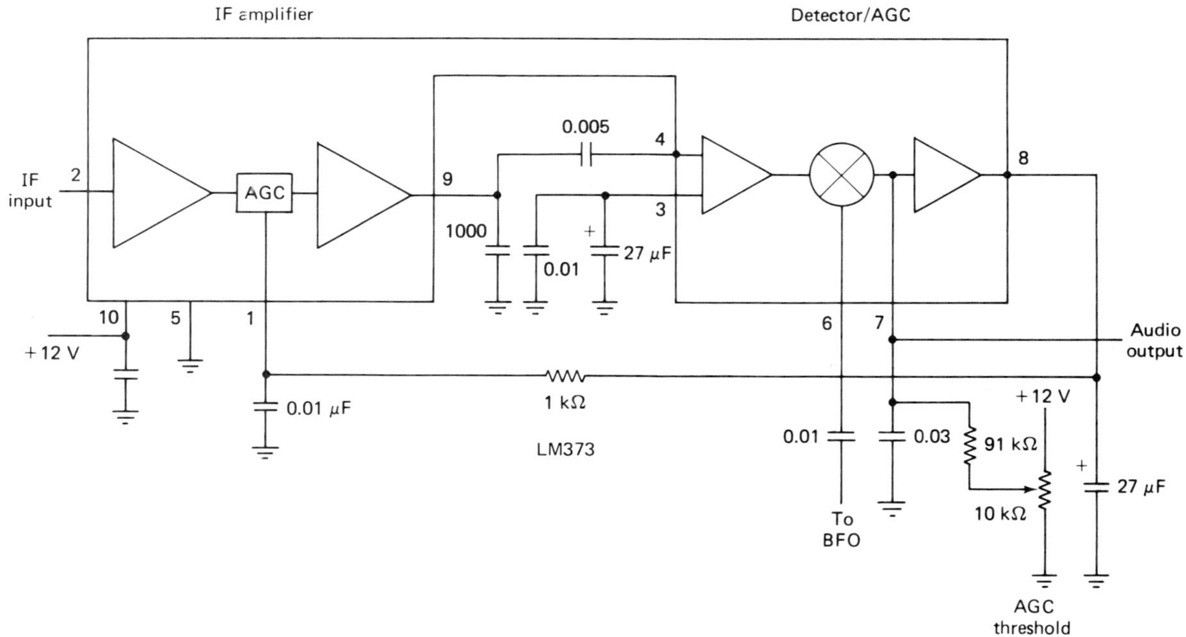

FIGURE 1-19. Complete 455-kHz IF amplifier, demodulator, and AGC system on a single chip.

radiation from the antenna, and any IF or BFO signals that leak into the audio amplifier are easily bypassed.

Double-balanced diode mixers make excellent product detectors with very low distortion. Designing around the relatively low impedance levels of the diode mixer is difficult, and as always, the main disadvantage of any diode mixer is the 6 dB or so of signal loss.

There are several integrated circuits that contain almost an entire IF system, including the demodulator, an AGC system, and several stages of gain. Most of the multipurpose IF ICs are for high-volume production items such as AM and FM broadcast radio receivers. There are several ICs suitable for serious communications receivers, one of which is made specifically for communications receivers and is shown in Figure 1-19 connected as a CW or SSB demodulator. This IC is capable of demodulating CW, SSB, AM, and narrow-band FM, depending on how the circuit is wired.

1.3.5 Audio Amplifiers

The audio-amplifier requirements for a communications receiver are not difficult to meet with common ICs. A power level of 1 W is sufficient for speakers, whereas 100 milliwatts (mW) is sufficient for headphone use. Since all high-frequency communications are narrow-band in nature, the bandwidth of the audio channel need be only a few kilohertz. Distortion is not critical and 1 percent or better is adequate. There are so many demanding applications for audio-amplifier ICs that there is a large selection of amplifiers suitable for communications receivers. A single IC audio amplifier suitable for headphone or speaker use is shown in Figure 1-20.

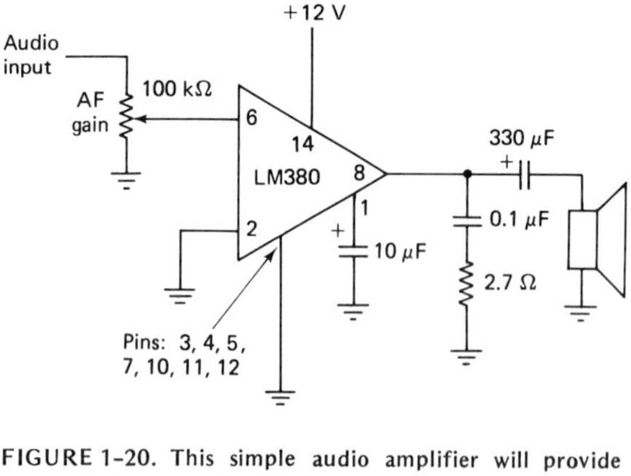

FIGURE 1-20. This simple audio amplifier will provide more gain than most demodulators require.

The audio amplifier in a direct-conversion receiver is a critical component because the entire gain of the receiver is obtained at audio frequencies. The noise figure of the audio amplifier in a direct-conversion receiver will determine, to a great extent, how good the noise performance of the overall receiver will be. Audio amplifiers suitable for use in the direct-conversion receiver may be adapted from tape recorder preamplifiers. The signal levels at the head of a magnetic tape recorder are very small and the requirement for the signal-to-noise level is strict. These amplifiers are for very specific applications and for specific values of gain. In some cases the amplifier is internally frequency-compensated for certain specific gain values which are not suitable for use in active filters or for gain values less than a certain minimum.

1.3.6 AGC Systems

Although an AGC system is a part of every communications receiver, there is practically no standardization in design because of the wide variation of circuits in a typical communications receiver. There are two basic types of systems in use: audio-derived and carrier-derived AGC.

Carrier-operated AGC systems are mostly for AM operation and were the mainstay of older communications receivers, where they were referred to as *automatic volume control* (AVC). In this system, the gain of the receiver is a function of the carrier level. In the case of the SSB signal, there is no carrier, and in addition, energy from the BFO can leak into the IF amplifier or AGC system and cause undesirable reductions in gain. Earlier communications receivers using the carrier-activated AGC often disabled the system when the BFO was operating and required the operator to control the gain of the receiver manually.

Figure 1-21 shows a carrier-operated AGC system. RF energy from the IF amplifier is rectified in D1 and amplified by A1 and applied to the gain-controlled stages. The output voltage in this system goes lower for increasing signal and is suitable for dual-gate MOSFET control. Bipolar transistors operating in the so-called reverse AGC mode and many integrated-circuit IF amplifiers require the reverse polarity, which can be created by reversing the diode, D1.

This AGC system has a dual time constant which is required for smooth AGC operation and the elimination of AGC "pumping." The amount of time required for an AGC system to act on an increase in signal level is called the *attack time* and should be short relative to the time required for the AGC system to recover from a reduction of the signal, or *decay time*. The AGC hold capacitor is charged through R1

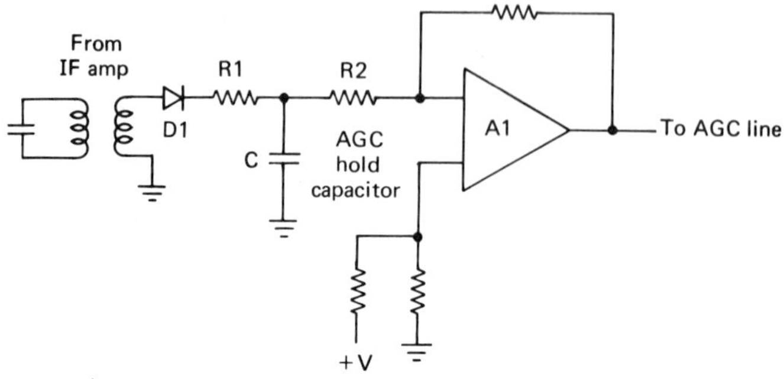

FIGURE 1-21. Carrier-derived AGC system resembles the simple diode detector used for AM detection.

and discharged through R2, where R1 is made much smaller than R2 so that the attack time is much shorter than the decay time.

It is important to isolate the BFO from the input of the AGC detector. This was often accomplished by picking off the IF signal for the AGC system well ahead of the product detector and supplying additional IF gain just for the AGC system.

The audio-derived AGC system is superior for SSB or CW reception but is not usable on AM. The audio system is similar to the carrier-operated system except that audio signals are rectified and charge the AGC hold capacitor as shown in Figure 1-22. Similar time constants are required in the audio system and are set by charging C through R1 for the attack-time constant and discharging through R2 to set the decay-time constant. As before, R1 is usually much smaller than R2.

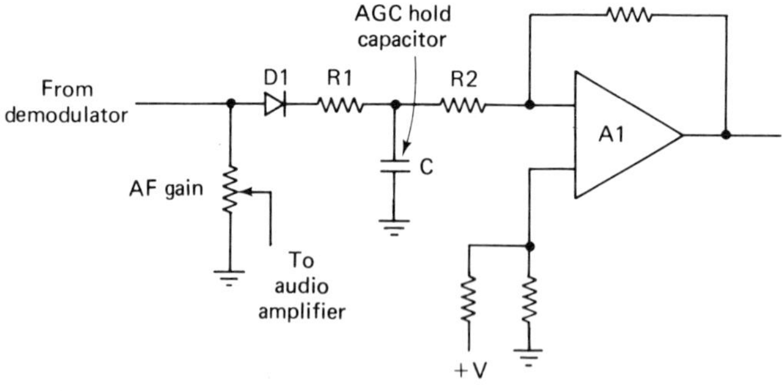

FIGURE 1-22. An audio-derived AGC system is superior for SSB and CW but does not function correctly on AM.

1.3.7 IF Amplifiers

The main purpose of the IF amplifier is to supply gain. Unlike the RF amplifier, which requires a good noise figure, low intermodulation distortion, and a wide dynamic range, the IF amplifier does not set the noise performance of the receiver and is fed signals that have had some gain control action and have been filtered by the IF filter. Therefore, the noise figure requirements, intermodulation distortion, and the dynamic range of the IF amplifier is much less critical than the RF amplifier. On the other hand, most RF amplifiers have less than 10 to 15 dB of gain, whereas the typical IF amplifier has over 60 dB of gain.

IF amplifiers are of two basic types, distributed gain and lumped gain. The *distributed IF amplifier* has several stages, usually using transistors for gain and coupled with either double-tuned transformers or simple single-tuned circuits. The *lumped IF amplifier* has only one gain element, an integrated circuit, and has no interstage coupling. In this type of amplifier, all of the IF selectivity must be obtained before the lumped-gain element, whereas in the distributed amplifier, the IF selectivity may be placed between the individual stages.

Figure 1-23 shows a two-stage distributed-gain IF amplifier using a double-tuned interstage transformer. In this amplifier the selectivity is obtained from all three IF transformers: the input, which is double-tuned; the interstage, which is double-tuned; and the output, which is single-tuned. The amount of selectivity that can be obtained from an IF amplifier of this sort depends mainly on the frequency of the IF. With the exception of very low IFs, the selectivity is inferior to a crystal or mechanical filter, and thus this circuit is used only in inexpensive receivers.

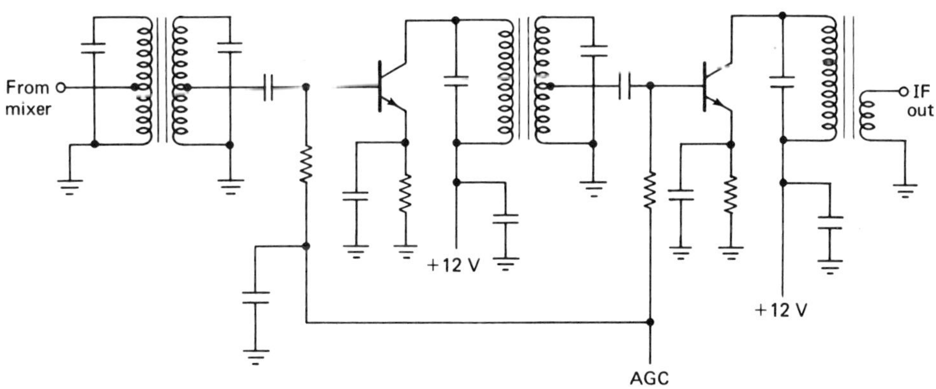

FIGURE 1-23. Two-stage IF amplifier using transformer coupling and AGC.

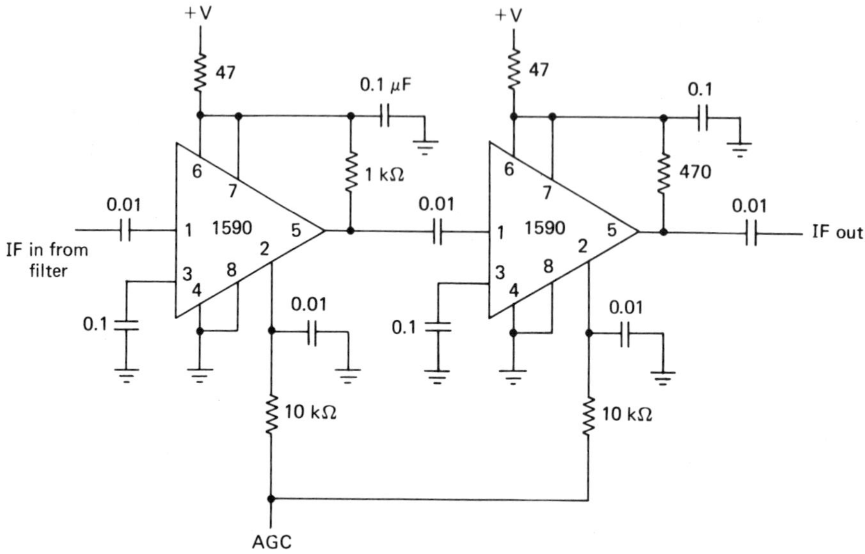

FIGURE 1-24. A wide-band untuned IF amplifier relies on a crystal filter for all its selectivity.

The lumped-gain IF amplifier shown in Figure 1-24 has only the input filter for selectivity and has no tuning of its own. There are many very high performance IF amplifier integrated circuits available suitable for communications receiver use that offer high gain and excellent AGC range. When using this sort of amplifier, the printed-circuit-board layout is very critical, because even the slightest feedback can cause oscillations.

1.3.8 Oscillators

Oscillators are used throughout the receiver for local oscillators, beat-frequency oscillators, and as reference oscillators for synthesizers, counters, and so on. Both crystal-controlled and *LC* oscillators are used throughout receiver designs.

A typical crystal-controlled oscillator used as an upper and lower sideband beat-frequency oscillator for a 455-kHz IF is shown in Figure 1-25. This oscillator is switched using dc from the front panel through the diodes D1 and D2. The modified Pierce circuit requires no tuning and is very stable.

A crystal oscillator for much higher frequency overtone crystals,

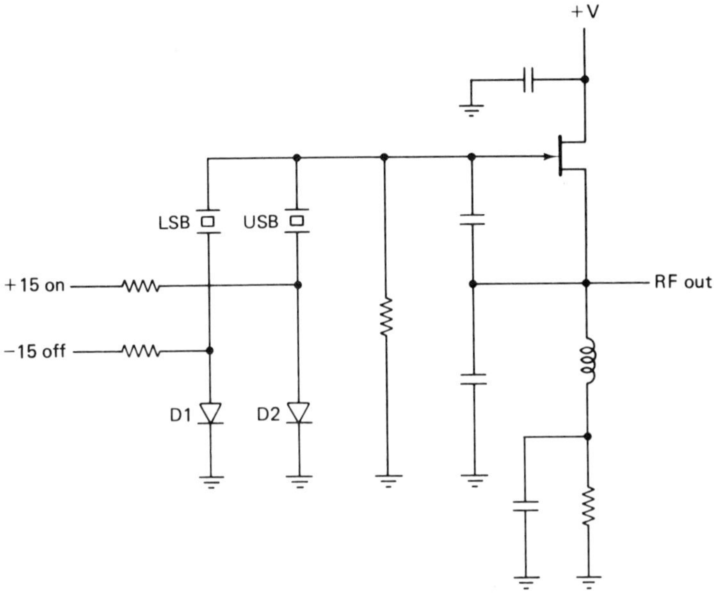

FIGURE 1-25. Electronically switched two-frequency oscillator.

such as would be found as the first local oscillator in a dual-conversion superheterodyne with a variable second IF, is shown in Figure 1-26. This circuit is suitable for use with third, fifth, and seventh overtone crystals up to 100 MHz.

Any of the standard oscillator circuits may be used as a tunable local oscillator or beat-frequency oscillator. One of the more interesting circuits is the Vackar oscillator shown in Figure 1-27. This oscillator has a steady output level over a wide tuning range and can be made

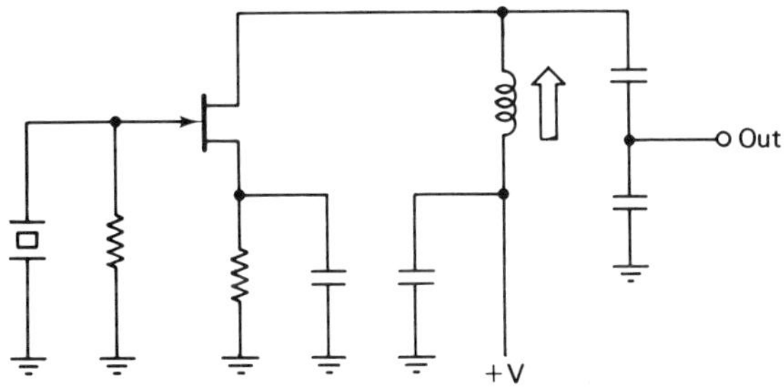

FIGURE 1-26. Crystal oscillator for overtone crystals. This circuit can operate with third, fifth, and seventh overtone crystals.

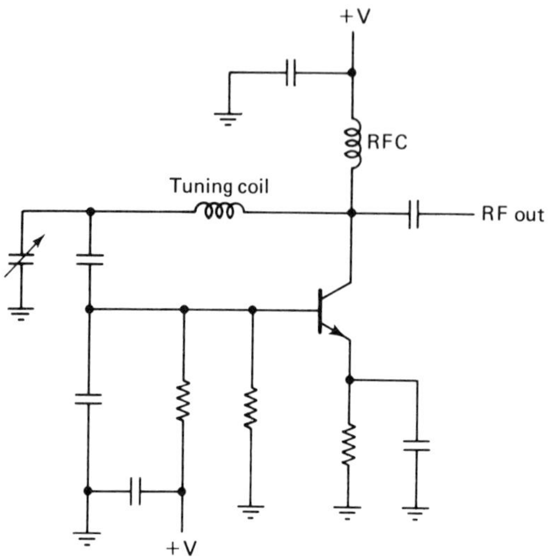

FIGURE 1-27. The Vackar oscillator makes an excellent local oscillator.

quite frequency-stable. Since the overall stability of a receiver is a function of all its local oscillators, the frequency drift of any local oscillator or beat oscillator appears in the overall receiver drift performance. Long-term temperature-induced drifts are compensated by using temperature-compensating capacitors in the local oscillator, while rapid small variations in frequency, known as *phase noise*, and mechanically induced frequency variations, called *microphonics*, are minimized by using high-Q tuned circuits mounted in rigid enclosures. In some cases, a buffer amplifier isolates the local oscillator from the mixer to decrease the effects of strong input signals on the local oscillator frequency.

Figure 1–28 shows two more conventional local oscillators suitable for high-frequency communications receivers. Figure 1–28a shows a Hartley oscillator without a buffer amplifier used for a dual-gate MOSFET mixer. Frequency pulling is kept to a minimum by tapping the oscillator at a low impedance point and feeding the second gate of a MOSFET mixer which has good isolation characteristics.

A Colpitts local oscillator that is used to drive a balanced mixer is shown in Figure 1–28b. In this circuit a FET amplifier is used to drive an NPN common-emitter buffer. The output is capable of driving a low-impedance mixer. The oscillator circuit operates from a regulated 9-V supply and has temperature-compensating capacitors to correct for frequency drift.

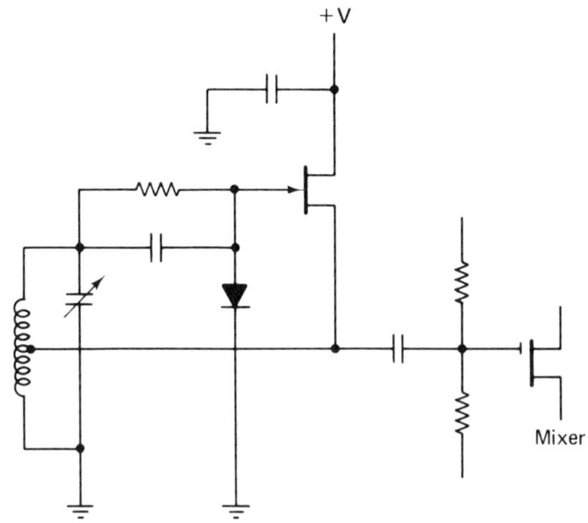

(a)

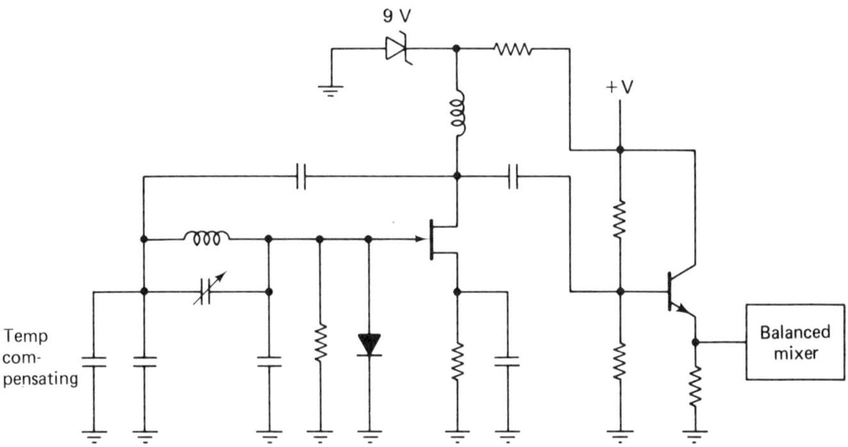

(b)

FIGURE 1-28. (a) Hartley oscillator driving a dual-gate FET mixer; (b) a Colpitts oscillator suitable for a local oscillator.

1.3.9 Accessory Units

Some accessory circuits, although not essential, have improved receiver performance. The S-meter has earned its place on the front panel of nearly every communications receiver. Originally supplied on receivers to aid in tuning AM signals, today the S-meter is used mostly for providing an indication of relative field strength.

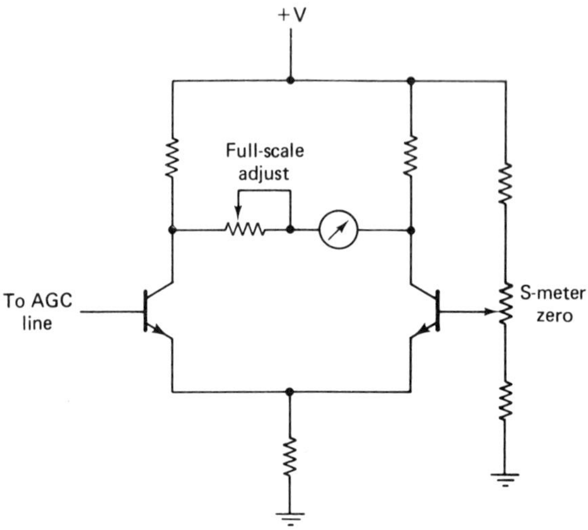

FIGURE 1-29. An S-meter amplifier is suitable for both normal and reverse AGC systems.

The S-meter is often nothing more than a voltmeter hooked to the AGC control line. Figure 1-29 shows a simple S-meter circuit, which can be used for the "normal" or "reverse" system by simply changing the connections to the meter. The S-meter is usually calibrated in arbitrary S units from 1 to 9, and in decibels above S9.

Passband tuning shifts a received signal within the bandpass of the IF filter without changing the pitch of a CW signal or without detuning a SSB signal. On a crowded band, the desired signal can be moved to the upper or lower edge of the IF filter while an interfering signal is moved completely out of the passband and no longer heard. Figure 1-30 shows graphically how an interfering signal is eliminated. A CW signal (a) is centered in an IF passband with a strong, interfering signal slightly above it. Adjusting the passband tuning control moves the de-

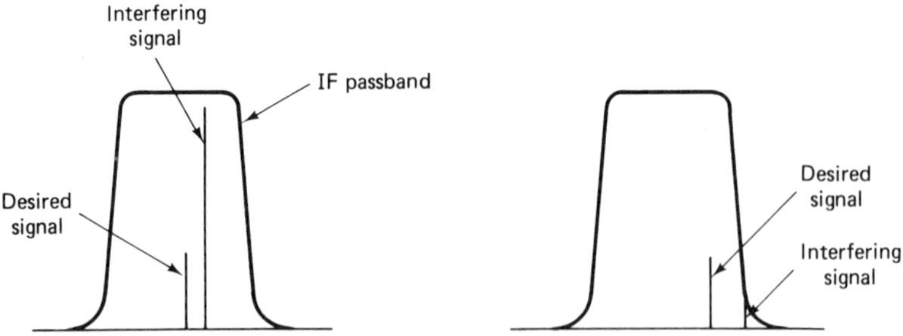

FIGURE 1-30. How passband tuning can eliminate interfering signals.

sired signal to the upper edge of the IF filter, where it can still be received. The interfering signal, however, moves beyond the passband and is no longer heard. The sharper the skirt response of an IF filter, the more effective will be its passband tuning action. The main tuning control of a receiver also moves signals within an IF passband. The pitch, however, also varies, sometimes to uncomfortably high or low frequency. To restore the pitch to its original frequency, the BFO must be adjusted. A passband tuning system automatically adjusts the BFO to maintain a constant pitch while the local oscillator changes slightly in frequency.

Passband tuning is difficult to achieve without causing spurious signals to appear in the receiver. Figure 1-31 shows the block diagram of a passband tuning system applied to a 48.05-MHz IF amplifier used in a sophisticated up-converting receiver. In this system, the second local oscillator and the BFO are both tuned with the passband tuning control while the first local oscillator remains constant. The second local oscillator and the BFO are both generated by heterodyning a 13.695-MHz voltage-controlled crystal oscillator with two crystals, one at 40 MHz to produce the local oscillator at 53.695 MHz and the other crystal at 8.05 MHz. Selection of the 13.695-MHz frequency was de-

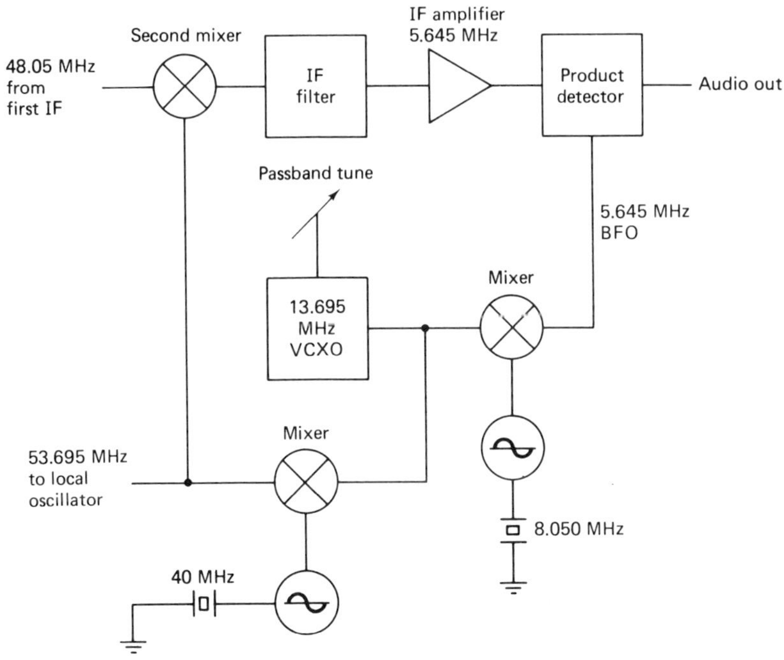

FIGURE 1-31. Passband tuning circuits from the Drake TR7.

termined by minimizing the spurious responses in the receiver. The frequency is chosen so that its harmonics and combinations of harmonics with other signals within the receiver do not fall within the frequency of the first or second IF. In most cases it is nearly impossible to achieve a totally spurious-free receiver. It is necessary to minimize the spurious signals and lower the level of all spurious signals to a level near that of the weakest input signal.

Another useful tool for eliminating interfering signals is the notch filter, which may be used for SSB as well as CW operation. The notch filter is essentially a very sharp band-reject filter that can be moved throughout the IF passband. The same two signals as in Figure 1-30 are shown in Figure 1-32, but the interfering signal is greatly attenuated by a notch filter. Notch filters are effective only if the notch is very narrow compared to the voice passband, so that the SSB signal is not destroyed. Such narrow circuits are difficult to build at higher IFs and are usually used at 455 kHz and lower. The schematic of a typical *Q*-multiplier type of notch filter for 455 kHz appears in Figure 1-33.

There has been a growing trend to replace the mechanical dial in communications equipment with a digital frequency readout. Because the local oscillator frequency in a communications receiver is not the same as the incoming received signal but is offset by the IF, the frequency counter used must have the capability to manipulate the counted frequency. The arithmetic is simple: the IF must be either added or subtracted from the counted frequency. For multiple-conversion superheterodynes, it may be necessary to count more than one local oscillator and manipulate several numbers. In a direct-conversion receiver, no manipulating is necessary because the local oscillator frequency is also the received signal frequency.

The calculations required to display the received frequency from

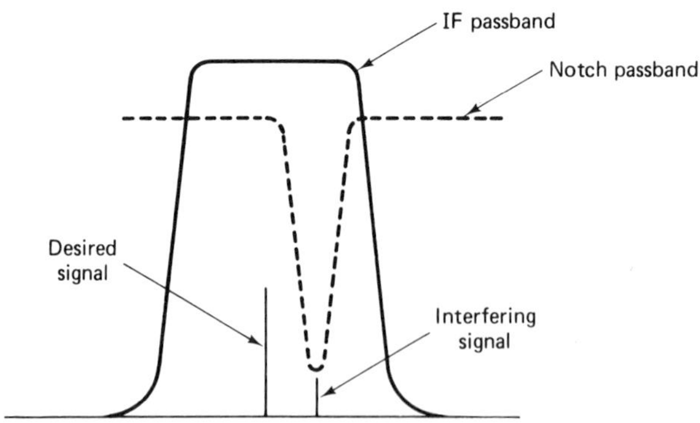

FIGURE 1-32. Notch filter being used to eliminate an interfering signal.

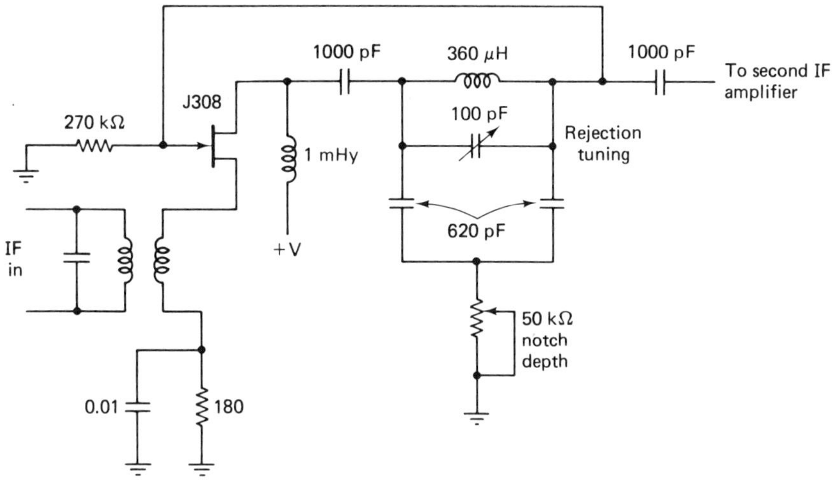

FIGURE 1-33. Notch filter for a 455-kHz IF.

the frequency counted can be accomplished in several ways. A direct approach is to count the required local oscillators and apply the data to a microcomputer to perform the arithmetic. If the receiver is a multiple-conversion receiver with passband tuning, this may be the only method. However, if the receiver is less complex, an indirect but less expensive method may be employed. One such method is to use an up/down counter and preset counters. If an IF is simply added to the counted frequency, a preset counter is used and before the counter is gated to count, the IF frequency is loaded into the counter. Instead of starting at zero, the counter starts at the IF frequency, which effectively adds the IF to any frequency counted. If the receiver is a multiple-conversion type, the counter is gated to count the first local oscillator, and then without clearing the count, is gated to count the second local oscillator. If necessary, one oscillator can be used to count up and the other to count down, which subtracts one local oscillator frequency from the other.

Another method for generating the correct digital readout from a frequency counter is to mix the local oscillators and beat-frequency oscillators and filter out either the sum or difference frequency as necessary. This method must be accomplished with extreme care so that the sum and difference frequencies generated for the frequency counter do not appear as spurious signals in the receiver.

In addition to valid radio signals, receivers are also subject to receiving noise. One type of noise that is particularly troublesome is impulse noise generated by automobile ignitions, electric motors, and other appliances. This type of noise is often strongest on the higher-frequency bands, such as 10 or 15 m. Since the energy of the impulse is spread out over a very wide frequency range, including the frequency to which the receiver is tuned, no IF or RF filtering of any sort can completely remove the noise energy. Two systems are in use to alleviate the nuisance caused by this noise. The simplest and most popular is the noise limiter shown schematically in Figure 1-34. This circuit reduces the amplitude of the detected noise spike rather than eliminating it. In this case, diodes D1 and D2 conduct when the detected audio signal reaches several hundred millivolts. Normal audio levels are kept below this level, with an effective AGC to prevent the diodes from clipping the audio. Impulse noise, because of its low-duty cycle, does not have a great effect on the AGC operation, but extremely high noise levels can cause the AGC to cut back on the gain.

A more effective but much more complicated system is the noise blanker. In this system, the impulse noise is actually eliminated rather than merely reduced. In this system, a complete AM receiver, as shown in the block diagram of Figure 1-35, operating at about 50 MHz, is coupled to the communications receiver antenna. The only purpose of this auxiliary receiver is to amplify and detect the noise signal. The detected noise is then used to temporarily disable the main receiver, thus preventing the noise impulse from being demodulated. Since impulses are narrow pulses, the short breaks caused by the noise blanker in the received audio do not seriously degrade the intelligibility. If the frequency spread of the noise signal does not include 50 MHz

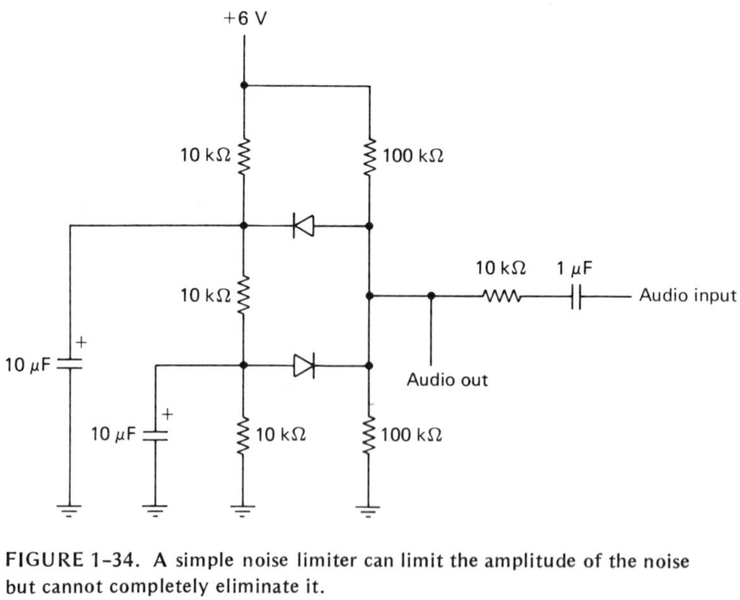

FIGURE 1-34. A simple noise limiter can limit the amplitude of the noise but cannot completely eliminate it.

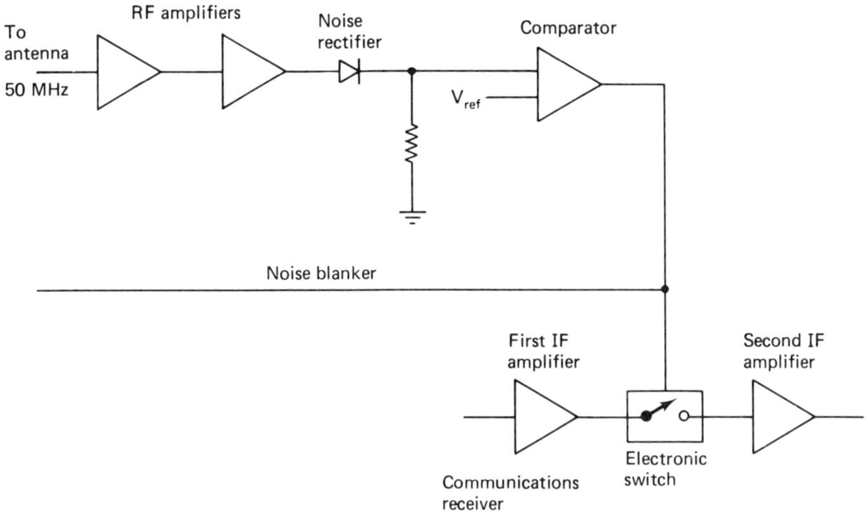

FIGURE 1-35. Noise blanker connected to a superheterodyne receiver.

(the input frequency of the auxiliary receiver), the noise blanker will not function. For most noise generated by motors, especially ignition noise, 50 MHz is an optimum frequency. Noises that are not eliminated by this system are atmospheric-generated noises, such as lightning and interference from low-frequency sources such as LORAN shore stations.

To ensure dial accuracy, high-quality receivers were equipped with a crystal calibrator that would provide accurate frequency markers every 100 kHz. The calibrator was nothing more than a crystal oscillator that had a very high level of drive to produce harmonics. The expense of the 100-kHz crystal and the development of digital logic has altered the design of crystal calibrators. A 1- or 2-MHz crystal is divided down with a flip-flop or counters to 100 kHz, 50 kHz, and 25 kHz. The digital signals are rich in harmonics and can be heard throughout the high-frequency band. A crystal calibrator using transitor-transitor-logic (TTL) integrated circuits is shown in Figure 1-36.

Following the discussion of many different receiver types (up-converting, single conversion, direct conversion, etc.), different amplifiers, mixers, oscillators, and so on, one may be confused as to just what makes a good receiver. Is there some magical combination of circuits into the right type of receiver that produces the superior communications receiver? The answer is absolutely not. A good communications receiver can be designed around almost any one of the types shown and with practically any combination of mixer, amplifiers, and demodulators. Of course, using the cheapest filter with the simplest

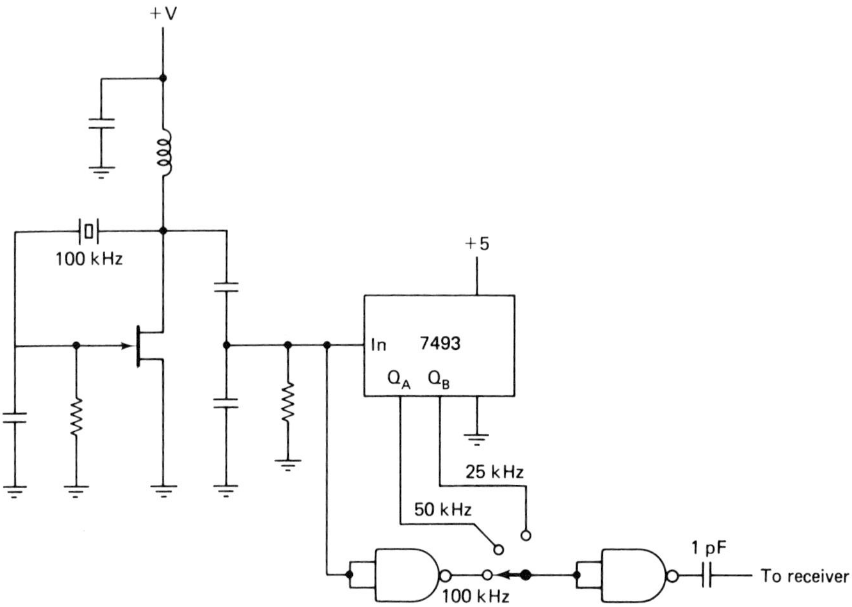

FIGURE 1-36. Dividing the crystal frequency in this crystal calibrator produces several marker frequencies.

mixer and the poorest-performing RF amplifier will not produce a high-performance communications receiver for high-frequency SSB reception. It may, however, produce a perfectly fine inexpensive receiver for listening to shortwave broadcasts at low cost. Then, is there a magical combination for amateur radio communications receivers? Not really magical, but two basic receiver types provide excellent performance for amateur communications. They are the single-conversion superheterodyne with an IF of at least 5 MHz and the up-converting superheterodyne. Either receiver type should be supplied with a phase-locked local oscillator with digital readout. An RF amplifier should not be used unless it uses a low-distortion amplifier with a very high third-order intercept. The mixer should be of the diode balanced type or a balanced mixer using matched FETs that have high-signal-level capability. Selectivity should be from crystal or mechanical filters and should under no circumstances be from an audio filter. There should be available some operating conveniences, such as passband tuning or a notch filter.

A receiver meeting all the criteria is not an inexpensive receiver circuit but does answer the question: What constitutes a good receiver? This does not imply that other combinations cannot produce a good receiver. Moderately priced and inexpensive amateur receivers cannot meet all the requirements, and some compromises must be made to bring the cost into line.

1.4 RECEIVER SPECIFICATIONS

Regardless of the receiver type—inexpensive, deluxe, large, or small—the capabilities of the receiver can be thoroughly explained through the use of specifications. To completely describe what may be expected of a communications receiver under all possible situations requires a complete set of specifications. Some describe the behavior of a receiver under weak-signal conditions, and other specifications describe the behavior of the receiver in the presence of strong signals. Others describe the characteristics of the receiver for both strong and weak signals together.

Receiver sensitivity, selectivity, and noise figure are three very important parameters that are interrelated. This is because noise is everywhere. Noise accompanies a radio signal from the transmitter through its journey through the atmosphere, in the receiving antenna, and even in the receiver itself. Noise is generated by the motion of charged particles and is an unavoidable part of nature. Noise is generated not only in the receiver and in the atmosphere, but a certain amount of noise even reaches the earth from outer space. The ratio of the signal to

The Health HR-1680 is a ham-band-only receiver available in kit form. (Courtesy of Heath Company.)

the accompanying noise is called the *signal-to-noise ratio* and is the one true figure of whether a radio signal will be useful or not.

Every amplifier or mixer adds some noise to a signal and hence degrades the signal-to-noise ratio. Amplifying a radio signal can never increase the signal-to-noise ratio of a radio signal. That means that if the signal-to-noise ratio of a signal were very poor, such that the signal was not intelligible, the best amplifier ever made could not be used to retrieve the signal. Therefore, the true ability of a receiver to use very weak signals with a low signal-to-noise ratio is dependent on the noise figure. The noise figure is an indication of how much damage is done to a signal by the receiver. For example, if a signal having a signal-to-noise ratio of 10 dB is passed through a receiver with a 3-dB noise figure, the signal-to-noise ratio at the output of the receiver would be only 7 dB. The higher the noise figure, the more noise that is added to the signal.

Noise has some characteristics different from signals. The amount of noise power received is dependent on the bandwidth of the receiver. Noise is not only everywhere; it is also all over the spectrum. Therefore, if a wide bandwidth is being received, a large noise power will be received. Reducing the bandwidth of a receiver will reduce the noise power received, and provided that the bandwidth is sufficient to pass the signal, the signal is unchanged. This will improve the signal-to-noise ratio of the signal. For example, if a CW signal were being received on an SSB filter and the signal-to-noise ratio were 3 dB (which would mean that the signal would barely be readable) and were received on a CW filter of 300 Hz, the signal-to-noise ratio would increase by 10 dB, to 13 dB. The sensitivity and noise figure of the receiver would remain unchanged and only the bandwidth would be changed to achieve the improved signal-to-noise ratio.

The sensitivity of a receiver is a measure of the weakest signal that can be received by a communications receiver. However, it was shown that the weakest signal capable of being received by a communications receiver depends on the noise figure and the bandwidth of the receiver. Therefore, the sensitivity of a receiver must be specified for a certain bandwidth and for a specific signal-to-noise ratio at the output of the receiver. Typically, the sensitivity is defined for a specific bandwidth, usually one that is optimized for SSB or CW communications. If the sensitivity were specified for both CW and SSB, it would be expected that the sensitivity would be greater for the narrower bandwidth for CW. A nomograph showing the relationship among noise figure, bandwidth, and sensitivity is shown in Figure 1–37.

In actual on-the-air operation in the high-frequency spectrum, the atmospheric noise is so much greater than any noise generated in the receiver that the noise figure of a modern communications receiver is seldom the limiting factor for sensitivity.

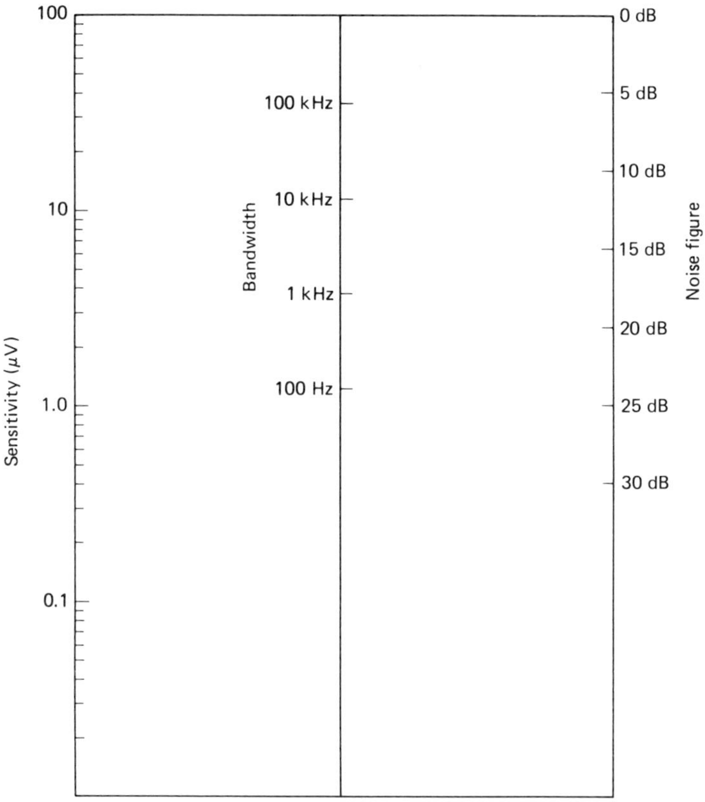

FIGURE 1–37. Sensitivity nomograph.

Selectivity is the ability of a receiver to reject nearby or adjacent channel signals while passing, undistorted, the desired channel. Two numbers are required to adequately specify selectivity; a 3- or 6-dB bandwidth and a shape factor. The shape factor is the ratio of the 6-dB to the 60-dB bandwidth.

In a modern communications receiver, the shape factor is more important than the actual 3-dB or 6-dB selectivity. Figure 1–38 shows how an SSB filter with a good shape factor is more valuable for CW than a poor CW filter with a much narrower 3-dB bandwidth. The passband of a good-quality SSB filter is shown superimposed on the passband of a simple CW filter. The single-sideband filter has a 6-dB bandwidth of 2.1 kHz and a shape factor of 1.5, whereas the CW filter has a 6-dB bandwidth of 500 Hz and a shape factor of 8. Notice that the attenuation of nearby signals is greater for the single-sideband filter beyond 1.6 kHz from the center frequency. It is evident that the wider

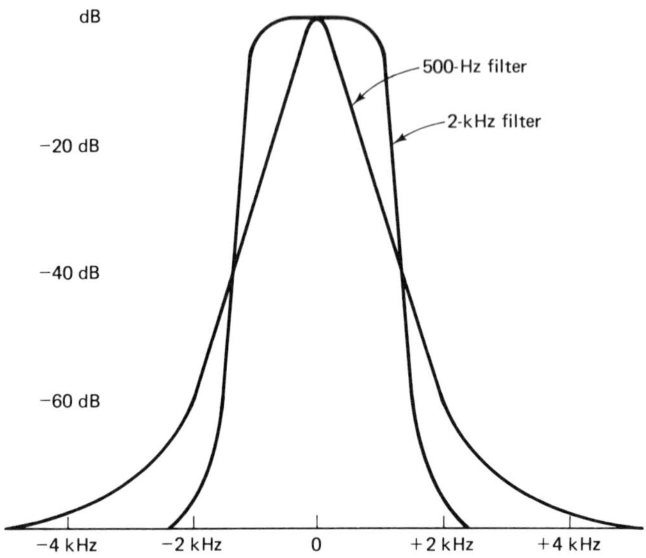

FIGURE 1-38. Effects of good skirt selectivity.

filter has more attenuation to signals 1.6 kHz removed, which indicates the extreme importance of shape factor.

Frequency accuracy and stability indicate a receiver's ability to be set to any desired frequency by use of the dial or digital readout and to "stay put" over time and temperature. Most receivers required a crystal calibrator to calibrate the dial to realize the specified dial accuracy, and the farther away from the crystal calibration point, the greater is the chance for dial error. Dial accuracy is usually specified as an absolute number and would most likely occur at a point midway between calibration points.

Frequency stability is given as the maximum frequency drift per unit of time, usually after a specified warm-up period. Occasionally, the drift is specified during this warm-up period. In the case of a mechanical dial, the drift can also contribute to the dial error. In a digital readout receiver, the counter will indicate the received frequency as it changes with time.

In some expensive receivers, the frequency change with temperature and power supply input will also be specified. This specification can be of importance, especially if the receiver is to be used for portable or mobile operation.

There are several specifications that can be lumped under the heading of spurious responses. A *spurious response* is a signal that can be heard where there should be no signal. These can be further divided into internally and externally generated.

An *internally generated spurious* is one that appears without any signal at the antenna input. The signals heard are harmonics, and mixing products of the internal oscillators of the receiver. These "birdies" and "whistles" are usually specified as no greater than so many microvolts. That is, a spurious signal will not be any stronger than a properly inputted signal of the specified microvolt level. It is particularly difficult to minimize the internally generated spurious in a communications receiver that uses multiple-conversion and/or digital readout. A good modern communications receiver can usually have spurious levels near or below that of atmospheric noise, which renders the spurious essentially unnoticeable.

An *externally generated spurious* requires an input signal before it can be heard. In some cases, two signals are required, and in other cases one signal can produce a spurious signal. *Image response* is an example of the latter type of spurious. Strong signals removed from the operating frequency by twice the IF may be heard. The elimination of the image is primarily a function of the front-end selectivity. Typical image rejections are in the range 60 to 70 dB.

Intermodulation and cross modulation are distortions produced by the RF amplifiers and mixer stages. Although distortions are produced by any stage, those ahead of the IF filter are most vulnerable.

Intermodulation is produced when two strong signals appear in the RF amplifier stage and are not only amplified but are mixed with differing combinations of harmonics to produce other unwanted signals. These spurious signals are referred to as third-order, fifth-order, and so on, intermodulation distortion. The order of the intermodulation is dependent on the mathematical expression describing the mixing action. The third-order distortion is the most troublesome in a communications receiver, and is usually the only one specified.

If two strong input signals are at frequencies of F_1 and F_2, the third-order intermodulation would appear at $2F_1 - F_2$ and $2F_2 - F_1$. For example, if two strong signals were at 7.05 and 7.1 MHz, the third-order intermodulation would appear at $2(7.05) - 7.1 = 7.0$ and $2(7.1) - 7.05 = 7.15$ MHz. A receiver with poor intermodulation performance tuned to either 7.0 or 7.15 MHz in the presence of those two strong signals would detect two spurious signals with possible modulation equivalent to the combination of the two signals. For example, if the strong signals were CW and SSB, the resultant intermodulation would be a distorted single-sideband signal keyed on and off with the CW signal.

There are several methods for specifying the third-order intermodulation performance. In older receivers, the issue was neglected entirely. One good way to specify the intermodulation performance is to use the concept of third-order intercept. Referring to Figure 1-39, any two

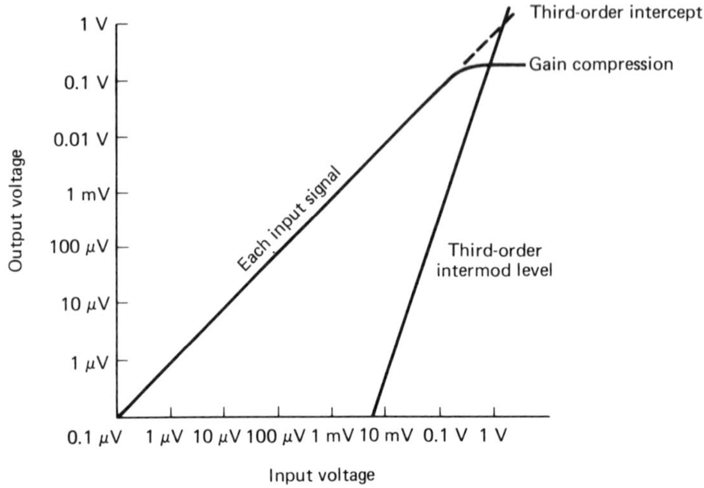

FIGURE 1–39.

input signals of equal level give rise to a third-order intermodulation product as shown in Figure 1–39. For every 1-dB decrease in the two input signals, the third-order product increases by 3 dB. Because of this difference in slopes, the desired signal line will intercept the third-order line at a point defined as the third-order intercept. Most mixers or amplifiers are incapable of reaching this level, making the intercept a theoretical point. It is at this point that the third-order product and the desired signals are equal in strength.

Another type of intermodulation is second-order or cross-modulation. This occurs when a very strong signal modulates a weaker signal within an RF amplifier or mixer. This is usually a phenomenon associated with amplitude modulation. For CW and SSB, the effect is not as great. However, a good source of cross-modulation are the very strong international broadcast stations found near amateur bands.

Occasionally, the performance of the AGC system in a receiver will be specified. This is usually tabulated as the largest change in input level that causes a 3-dB change in output level.

The specifications outlined above define the ability of a communications receiver to handle strong signals, weak signals, strong and weak signals together, and signals close together in frequency. Other specifications are useful for special requirements, such as frequency coverage, power supply requirements, and audio power output. Also, specifications pertaining to special features, such as notch filters and S-meters, are required.

2 HIGH-FREQUENCY TRANSMITTER AND TRANSCEIVER FUNDAMENTALS

2.1 SIMPLE CW TRANSMITTERS

A radio transmitter is required to perform one basic function: to generate a modulated radio-frequency carrier. More than anything else, the type of modulation dictates the type and complexity of a radio transmitter. Transmitters for simple on/off keying used for the transmission of Morse code are the simplest, whereas those for single sideband are the most complex. Three basic schemes make up the majority of amateur-band transmitters. They are the master oscillator-power amplifier, the frequency multiplier (and the more modern frequency-divider adaptation), and the heterodyning type. Curiously, as the superheterodyne receiver became the most popular receiver type, the heterodyning transmitter has become the only transmitter system in widespread use in new equipment designs. Although the superheterodyne receiver was accepted very early in the history of radio, the heterodyning transmitter required an additional 30 years or so to gain popularity.

The most elementary form of transmitter is the *master oscillator-power amplifier* (MOPA) *transmitter*, which appears in Figure 2-1. Although there are single-stage transmitters, that is, just an oscillator feeding an antenna directly, they suffer from very serious drawbacks,

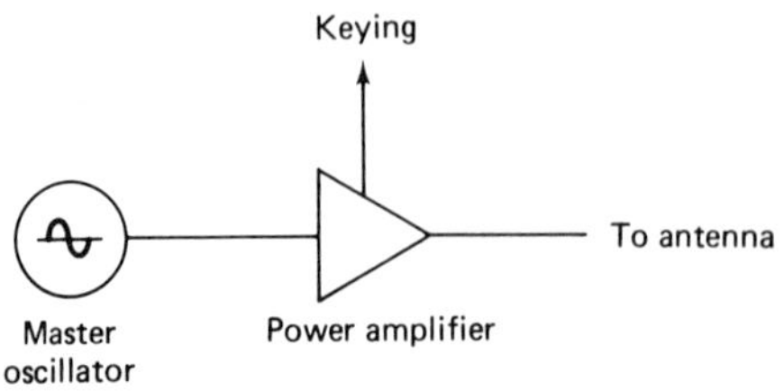

FIGURE 2-1. The master oscillator-power amplifier transmitter was a very popular circuit during the early days of radio.

are usually for crystal-controlled CW use only, and are not considered as a successful transmitter type.

The oscillator operates at the desired transmitter frequency and its relatively low power is amplified to the desired power by the RF amplifier. Additional amplifiers between the oscillator and the antenna can increase the isolation between the output and the oscillator, thus enhancing the frequency stability, or the power can be raised to a higher level than is possible with only two stages. Even with several stages of amplification, the transmitter falls into the MOPA category. It is desirable to have the oscillator operating at all times and to key the amplifier stages, to prevent rapid frequency drift, better known as *chirp.* However, the continuously operating oscillator can be heard in the station receiver and is a source of harmful interference. Turning the oscillator off when receiving can eliminate the interference problem but can introduce a frequency drift at the beginning of each transmission.

Another serious problem of this simple transmitter is that the tuned circuits for the entire transmitter must be changed for each band of operation. Transmitters of this kind tend to be single-band affairs for home building, having the advantage of low spurious outputs and ease of construction.

A solution to the interference problem plus some relief from the bandswitching problem is shown in Figure 2-2. In this design the master or *variable-frequency oscillator* (VFO) does not operate at the final frequency but at a fraction of that frequency while frequency multipliers raise the VFO up to the desired output. If the multipliers are keyed, as were the amplifiers in the MOPA transmitter, the VFO may be run continuously. During periods of receive, the VFO is operating at a frequency below the frequency to which the receiver is tuned. If care if taken not to allow excessive harmonics to be generated, and good shielding is used, the VFO will not be heard in the receiver.

Because the amateur bands are harmonically related, that is, the 80-m band is roughly half of the 40-m band frequency, the 40-m

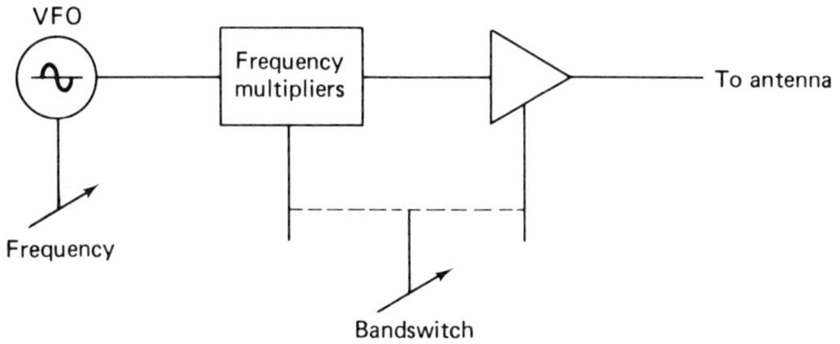

FIGURE 2-2. The frequency multiplier transmitter is capable of operating on several amateur bands because of their harmonic relationship.

band is half of the 20-m band frequency, and so on, the same VFO may be used for several amateur bands by simply multiplying its frequency with frequency multipler circuits.

A typical CW transmitter from a few years back would have a VFO operating in the 160-m band, which was running continuously. The VFO frequency would be doubled for 80-m operation, quadrupled for 40 m, multiplied by 8 for 20 m, multiplied by 12 for 15 m, and multiplied by 16 for 10 m.

This scheme suffers from a very serious problem. Not only is the VFO frequency multiplied, so is any frequency drift. If the VFO drifts 100 Hz, the amount of frequency change at 10 m is 1600 Hz. Similarly, the tuning rate would be different for each band. For example, if one revolution of the VFO knob moved the frequency 10 kHz on 80 m, which is a good tuning rate, the rate would be 80 kHz per revolution at 10 m, which is a bit fast. This makes precise frequency setting and dial accuracy difficult on the higher-frequency bands.

Another form of CW transmitter makes use of the harmonically related bands in order to use a single-range VFO at a high frequency and divides the frequency with digital logic to the amateur bands as shown in Figure 2-3. The master oscillator or VFO operates above all the amateur bands covered by the transmitter. Therefore, the VFO may be operated continuously since the fundamental and all the harmonics of the VFO frequency lie outside the amateur bands. Frequency division is accomplished by using high-speed digital logic such as *emitter-coupled logic* (ECL). These integrated circuits require no tuned circuits as do frequency multipliers, so that all the frequency-dividing circuits may be mounted on a printed circuit board without fear of instability. Bandswitching may be accomplished electronically by using ECL gates and multiplexers.

Several of the disadvantages of the multiplying type of CW transmitter apply to the divider type. The high frequency of the VFO makes

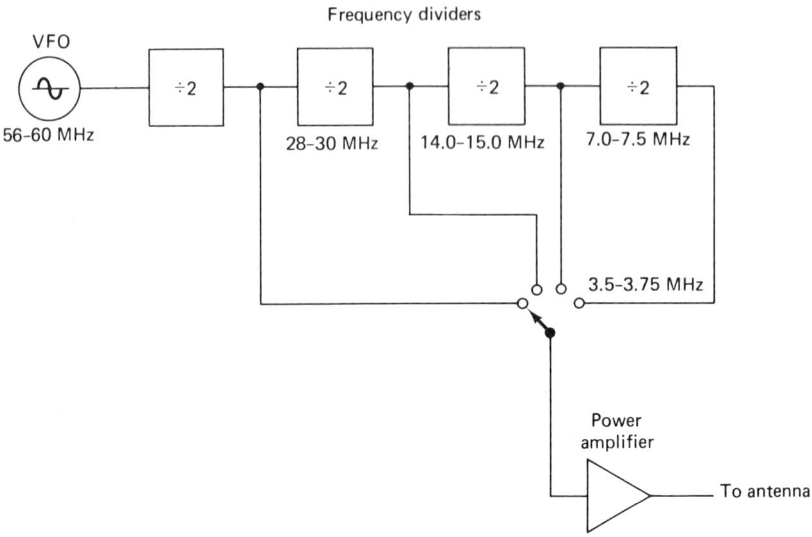

FIGURE 2-3. The frequency-divider type of transmitter is the modern equivalent to the frequency-multiplier type.

it difficult to achieve good stability. However, frequency division not only divides the frequency but also divides the frequency drift. If the VFO drifts, for example 1.6 kHz, the equivalent drift on the 80-m band is only 100 Hz. As it was in the multiplier transmitter, the worst band for frequency stability is 10 m. Similarly, the tuning rate would be different on each band. These problems are exactly the same as those encountered in the multiplier transmitters.

An excellent solution to the tuning-rate and frequency-drift problem is to heterodyne the VFO to the desired amateur band rather than multiply, as shown in Figure 2-4. A heterodyning transmitter is somewhat like a superheterodyne receiver in reverse. The VFO operates at a convenient frequency such as 5.0 to 5.5 MHz and is beat with a crystal oscillator to reach the desired amateur band. Since no frequency multiplication is involved. the tuning rate and frequency drift remain unchanged, regardless of the band of operation. Also, since the VFO is not operating at the final frequency or any fraction thereof, it may be operated continuously without any interference to the receiver.

A problem unique to heterodyning-type transmitters is *dial reversal.* If the output frequency is the sum of the crystal oscillator and the VFO frequencies, increasing the VFO will increase the frequency of the output. However, if the output frequency is the difference of the crystal oscillator minus the VFO, then increasing the VFO frequency decreases the output frequency. Although the tuning rate is unchanged, the direc-

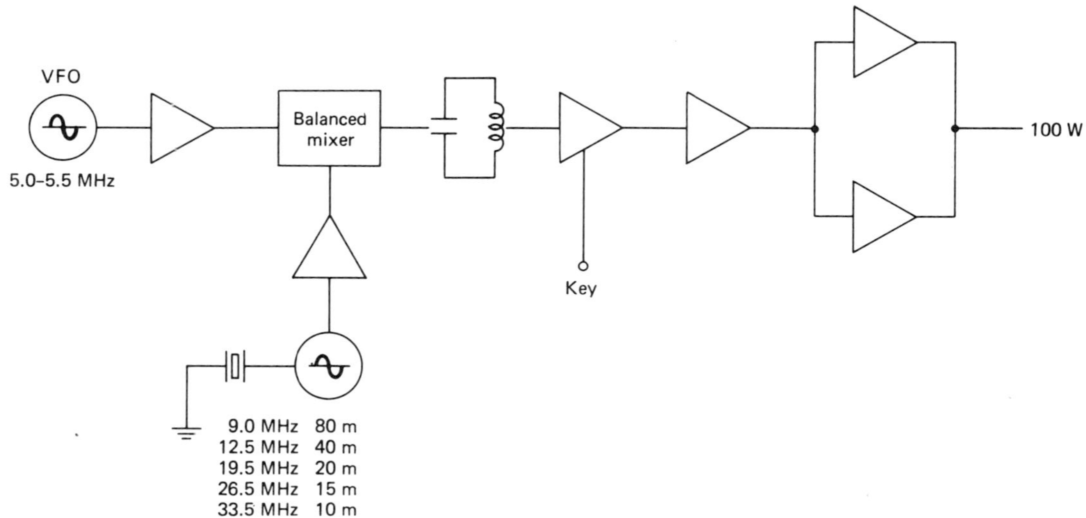

FIGURE 2-4. Block diagram of the Heath HX1681 CW transmitter.

tion has reversed. If a transmitter has bands that use both the sum and the difference, it will be necessary to include two dial calibrations, one for the forward tuning and one for the reverse.

Heterodyning transmitters play an important role in modern transmitter design, especially for SSB transmitters, where stability is an important requirement and frequency multiplication is not possible. Practically all modern high-frequency transmitters use the heterodyning principle, both amateur and commercial.

2.2 SINGLE-SIDEBAND TRANSMITTERS

The most common transmitter in use on the amateur bands is the single-sideband heterodyning transmitter. SSB is a very efficient type of modulation. It uses no more spectrum space than is required and makes efficient use of transmitter power. Single sideband is a specialized form of amplitude modulation where components of the amplitude-modulated signal that are not necessary for the transmission of information are removed. Either sideband of an amplitude-modulated signal contains the modulation information, while the carrier contains no information whatsoever. Thus, only one of the two sidebands need be transmitted.

2.2.1 Sideband Generators

There are several methods of generating a single-sideband signal, some practical and some only theoretical. Only two methods, the filter method and the phasing method, have been used in amateur transmitters.

Generating the SSB signal with the filter method is shown in Figure 2-5. This method starts with a double-sideband suppressed carrier signal, which is generated with a balanced modulator, and the undesired sideband is stripped off with a high-performance filter. The spectrum of the balanced modulator output as shown in Figure 2-6a contains both sidebands but no carrier. After passing through the sideband filter, one sideband is removed, leaving only a single sideband, as shown in Figure 2-6b.

The sideband filter is a high-performance filter with very sharp skirts. The typical voice-frequency spectrum is from 300 Hz to 3 kHz, which is a total bandwidth of 2.7 kHz. When a carrier is modulated with a typical voice-frequency modulation, the unwanted sideband is only 600 Hz to 6 kHz away. Therefore, to eliminate the unwanted

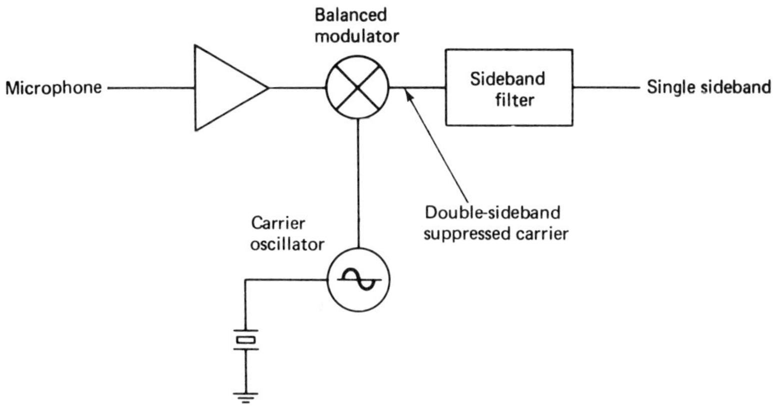

FIGURE 2-5. A single-sideband generator is required for every type of sideband transmitter or transceiver.

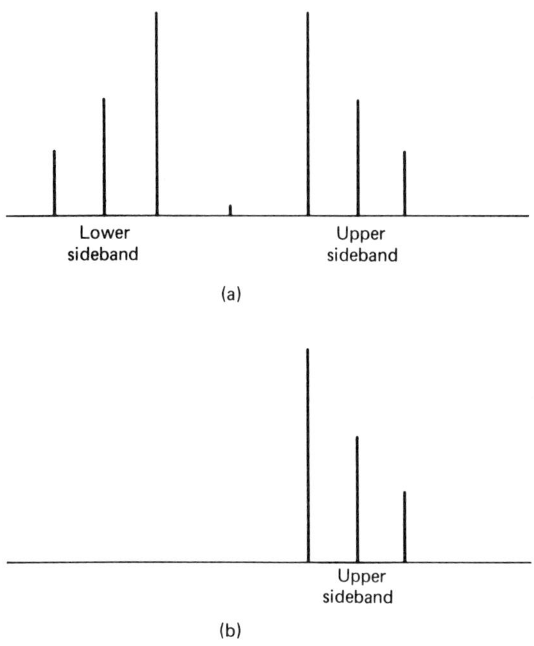

FIGURE 2-6. (a) A double-sideband spectrum contains both the upper and lower sideband but no carrier; (b) after passing through the sideband filter the double-sideband signal becomes single-sideband.

sideband while passing the desired sideband, a filter with a good shape factor is required. Mechanical filters are used for low-frequency sideband generators, such as 455 kHz, while crystal filters are also used for low frequencies and up to 10.7 MHz. The filters used in a single-sideband generator are exactly the same used in communications receivers. The similarity of transmitter and receiver filters and the very high cost of a good-quality filter led to the development of the SSB radio transceiver to share the filter between transmit and receive.

Placement of the carrier frequency in the filter passband determines whether an upper or a lower sideband will be generated. By placing the carrier 10 dB or so down on the filter skirt, an additional carrier suppression can be achieved without detracting from the audio-frequency response. The same sideband filter may be used to generate either the upper or lower sideband by changing the frequency of the carrier oscillator.

The phasing type of single-sideband generator was used primarily to avoid the very expensive sideband filter. In this generator, two balanced modulators are used and are fed from two carrier signals which are identical except for a 90-degree phase shift called *quadrature.* (If one signal is in quadrature with the other, it is exactly the same frequency and amplitude but 90 degrees out of phase.) Two audio signals are used that are exactly the same except for a 90-degree phase shift between all components of the audio waveform. In the phasing type of SSB generator, as shown in Figure 2-7, the output of each balanced modulator

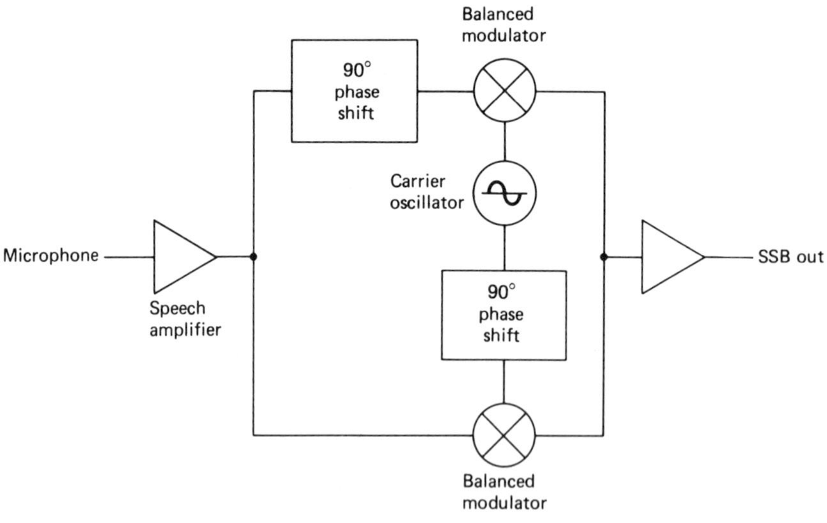

FIGURE 2-7. The phasing type of sideband generator does not require a sideband filter.

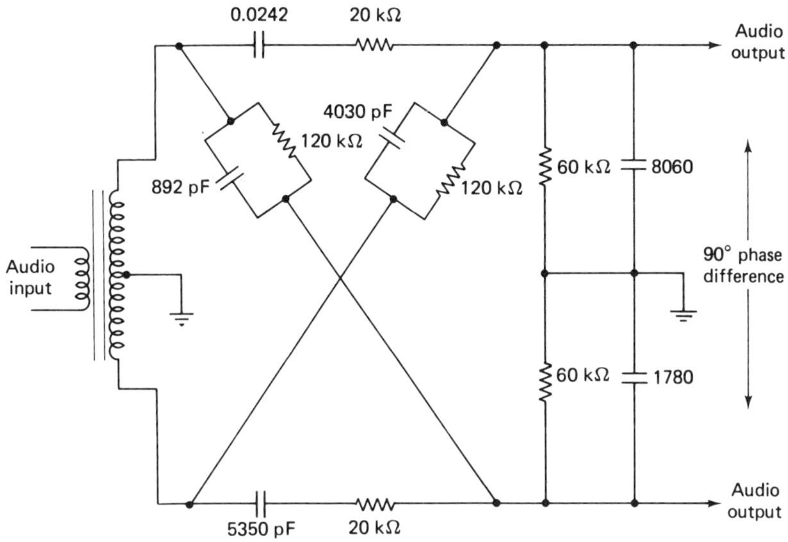

FIGURE 2-8. An audio-phase-shift network.

is a double-sideband suppressed carrier signal with a phase relationship such that if the two signals were added together, one of the sidebands would cancel and the result would be a single sideband. Which sideband remains depends on the phase relationship between the carrier or audio inputs to the balanced modulator.

The amount of carrier suppression depends on the suppression of each balanced mixer, and the unwanted sideband suppression is a function of the accuracy of the phase angle between the carrier and audio-frequency signals fed to the balanced modulators, and the accuracy of the summing of the outputs of the balanced mixer.

Creating an accurate 90-degree phase shift for all audio frequencies in the voice-frequency range is the single most difficult task in the design of the phasing-type exciter. Usually, the phase shift is generated with a very carefully designed and constructed *RC* network that is made with precision components of nonstandard values. One very popular phase-shift network is the Barker and Williamson 2Q4 shown in Figure 2-8, which is manufactured in a metal can much like an old vacuum tube. In this way, the components were shielded from hum pickup from local sources, thus assuring that all the components were at the same temperature for maximum stability.

Generating the RF quadrature is not difficult, especially if a single frequency or a narrow band of frequencies are required. A simple *RC* or *LC* network will provide a stable, accurate quadrature network.

Producing a quadrature signal over a very wide frequency range

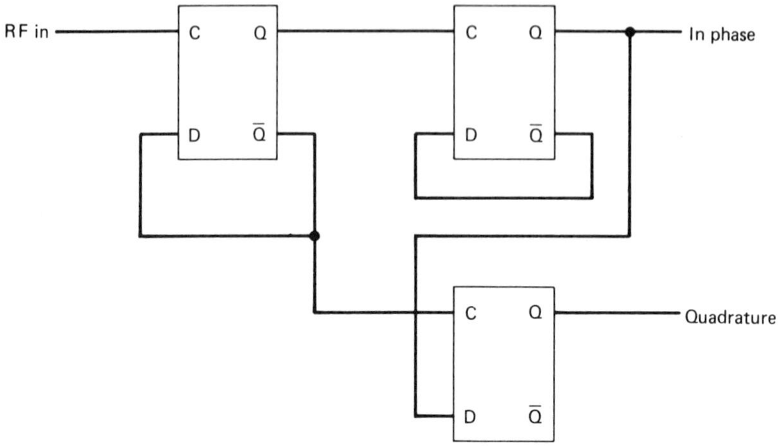

FIGURE 2-9. RF quadrature can be easily generated with digital logic.

must be accomplished by digital logic as shown in Figure 2-9. This system does not rely on the reactance of any component and is insensitive to frequency. A VFO operating at four times the desired carrier frequency and very fast digital logic is required for good unwanted sideband rejection. For frequencies in the 40-m band and lower, Schottky-clamped TTL will be sufficient, but for the 20-m band and higher, emitter-coupled logic (ECL) must be used.

The primary disadvantage of the digital quadrature generator is the rather high frequency required for the VFO. A digital quadrature generator operating in the 10-m band at 29 MHz would require a VFO at 116 MHz. Since the VFO frequency is divided by a factor of 4, any frequency drift will be reduced by a like factor. However, the 116-MHz oscillator requires very high speed logic and construction practices not usually found in high-frequency radio systems.

The significant advantage of the digital quadrature generator is that it allows the single sideband to be generated at the operating frequency rather than generating the SSB at a fixed frequency and heterodyning it up to the operating frequency.

Another method of sideband generation that has received occasional attention is the *Weaver*, or so-called *third method*. This is a potentially useful, special phasing type of sideband generator that uses a subcarrier.

The Weaver method bypasses the need to generate a wide-band audio phase-shift network by mixing the incoming audio signal with a 1.7-kHz subcarrier and using both the in-phase and the quadrature of the subcarrier, which can easily be generated with logic circuits. One of each mixing product is eliminated by the use of a low-pass filter, as

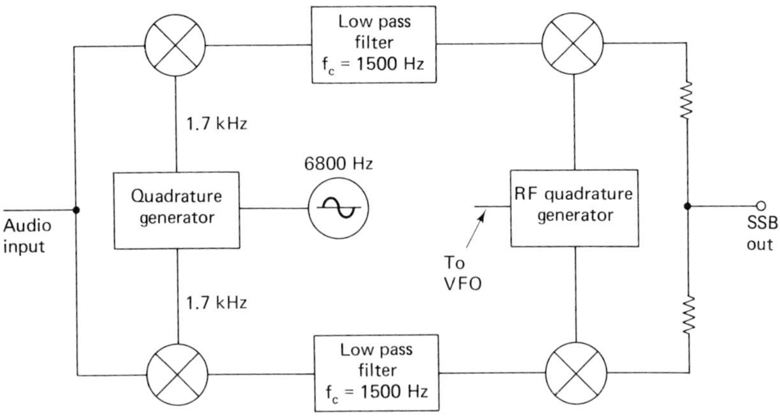

FIGURE 2-10. The Weaver, or "third method" of single-sideband generation.

shown in Figure 2-10. The outputs of the low-pass filters are the phase-shifted audio signals that may be applied to the dual balanced modulators as in any phasing-type generator.

2.2.2 Heterodyning Transmitters

The SSB signal, as it emerges from any of the sideband generators described, is fully modulated and no futher modulation is required. However, great care must be taken with the design and alignment of the stages in the transmitter so that no distortion or unwanted modulation is applied to the signal. That is, all the amplifiers and mixers in the single-sideband transmitter must be linear. The output of a linear stage has exactly the same waveshape as the input, and any deviations from this basic waveshape will cause distortion.

Unlike the CW signal, a single-sideband signal may not be multiplied to translate it to a desired amateur band; that is, all frequency translation must be done by mixing. Therefore, a block diagram of an SSB transmitter looks much like the block diagram of a superheterodyne receiver, with local oscillators, mixers, and IF amplifiers. This is another similarity of the SSB transmitter to a superheterodyne that makes it desirable to integrate transmitter and receiver into an SSB transceiver.

Figure 2-11 shows a block diagram of a single-conversion SSB transmitter in which two-band operation is accomplished by using both the sum and difference of the 9-MHz SSB signal and a 5.0- to 5.5-MHz VFO. The sum is 20 m and the difference is 75/80 m (3.5 to 4.0 MHz).

This scheme causes a dial reversal as discussed in the section on heterodyning-type CW transmitters. With the VFO set to 5.0 MHz, the

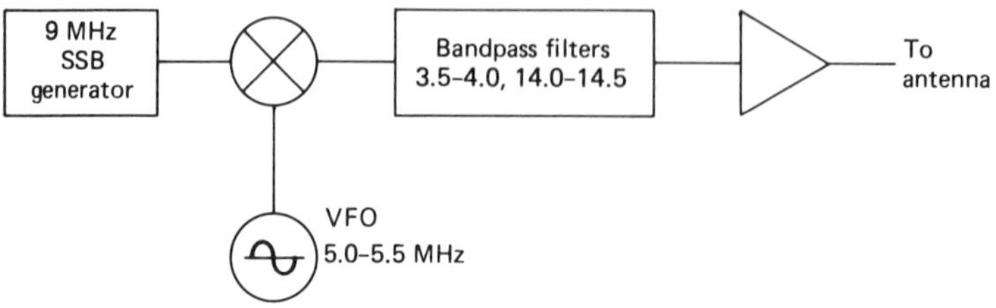

FIGURE 2-11. Single-conversion sideband transmitter that generates two amateur bands by using the sum and difference frequencies.

output frequency on 20 m is 14.0 MHz, and increasing the VFO to 5.5 MHz increases the output to 14.5 MHz, which is normal tuning. On the 75/80-m band, a VFO frequency of 5.5 MHz corresponds to 3.5 MHz and goes up with decreasing VFO frequency to 4 MHz, which corresponds to a VFO frequency 5.0 MHz, which is reverse tuning. The dial calibration in the normal case may go from 0 to 500, but in the reverse case the dial must be calibrated from 500 down to 0.

Like a communications receiver, the single-sideband transmitter has images. The band-pass filter following the mixer and between the amplifiers is for eliminating these images. As previously mentioned, all the mixers and amplifiers must be linear.

It is possible to cover more than two amateur bands with a single-conversion transmitter, but the high VFO frequency necessary for the higher-frequency bands places a very rigid frequency-stability requirement on the VFO.

Figure 2-12 shows a simple, single conversion SSB transmitter for all amateur bands. The local oscillator frequency is 14.5 to 15.5 MHz for the 80/75-m band, 16.0 to 16.5 MHz for the 40-m band, 11.0 to

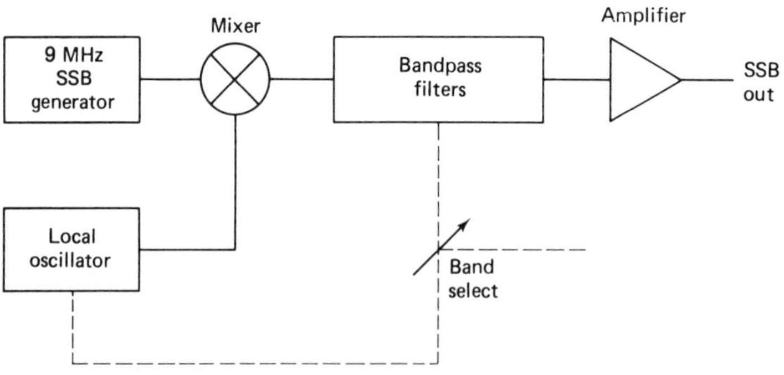

FIGURE 2-12. Single-conversion sideband transmitter using a multiple-band local oscillator.

11.5 MHz for the 15-m band, and 19.0 to 20.7 MHz for the 10-m band. Since the stability is better at lower frequencies, the 20-m band, because it has the lowest local oscillator frequency, will have the greatest stability. On the 80/75-m and 40-m bands, the output frequency is equal to $F_{OUT} = F_{VFO} - F_{IF}$; on all other bands the frequency is equal to $F_{OUT} = F_{VFO} + F_{IF}$. Notice that the term F_{IF} has a plus sign in one equation and a minus sign in the other. This indicates that there will be a sideband-inversion problem in this transmitter. This indicates that if the lower sideband is generated at 9 MHz, it becomes the upper sideband at 7 MHz, or 3.5 MHz, which is called *sideband inversion.* Translated to any of the other bands, there would be no inversion and the output would be lower sideband. A perfectly good SSB signal is produced at the output, but the upper and lower sideband nomenclature on the front panel would have to be reversed on 40 m and 80/75 m. This could be handled by either marking the front panel with the opposite sideband nomenclature for use only on 40 m and 80/75 m or actually switching the sideband generator internally.

A much more elegant way to generate an output on all the amateur bands without frequency-stability problems, sideband inversion, or dial reversal is the use of *dual conversion.* Dual conversion in the communications receiver provides improved image rejection from the higher-frequency first IF and good selectivity from the second, lower-frequency IF. In the SSB transmitter, image rejection and good unwanted sideband rejection are the benefits of dual conversion. Like the receiver, there are advantages to using single conversion in the SSB transmitter. The trouble arising from dual conversion is caused primarily by images and spurious signals, which can be easily attenuated to below the legal limit in the transmitter, but could be insurmountable in the receiver.

The SSB generator of the typical dual-conversion SSB transmitter in Figure 2-13 operates at 455 kHz using a mechanical filter for sideband generation. The 455-kHz frequency allows one of many standard sideband filters to be used, and because the frequency is relatively low, allows for good carrier suppression. A VFO operating from 2.7 MHz down to 2.5 MHz is heterodyned with the 455-kHz sideband signal to create the first IF, which spans the frequency of 3.155 down to 2.955. The dial of the VFO is calibrated in reverse; that is, the highest VFO frequency, 2.7 MHz, is calibrated as 0 and the lowest VFO frequency has the highest dial number, 200. Since all the amateur bands are generated by subtracting the first IF from a crystal oscillator, the dial direction will be correct at the output frequency. Mathematically, this may be written $F_{OUT} = F_{XTAL} - F_{IF}$, where F_{OUT} is the frequency of the output within the amateur band, F_{XTAL} is the frequency

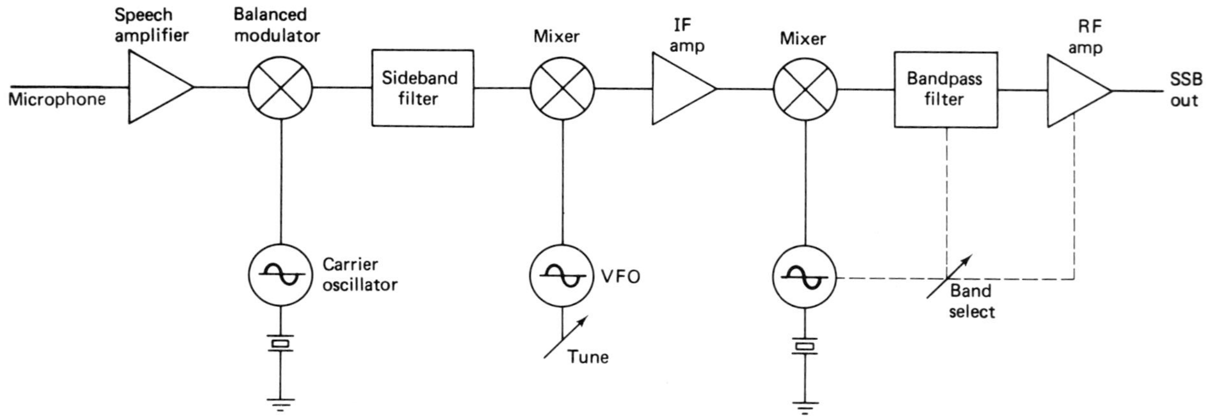

FIGURE 2-13. A dual-conversion SSB transmitter can operate a multitude of bands without frequency-stability problems.

of the crystal oscillator, and F_{IF} is the frequency of the first IF, 3.155 to 2.955 MHz. The minus sign in front of the second IF term indicates that not only is there a dial reversal but that there is a sideband reversal as well. These reversals occur on all the amateur bands, so once the dial is properly calibrated and the sideband nomenclature is placed on the front panel, it does not change from band to band.

Any 200-kHz band may be covered from 3.2 MHz to 30 MHz with the exception of 5.91 to 6.31 MHz, which is avoided because of the possibility that energy from the second harmonic of the second IF will be amplified and sent to the antenna. Since this frequency does not correspond to any amateur band, this system is capable of covering all amateur bands without significant spurious emissions.

Notice that one dial calibration is required to cover all possible output bands. The dial frequency, 0 to 200 kHz, is simply added to the frequency set by the band switch. For example, if the bandswitch (crystal oscillator) is set at the 3.5-MHz position and the VFO dial is set to 155, the output frequency is 3.655 MHz. Internally, the VFO frequency is 2.545 MHz, the second IF is 3.000 MHz, and the crystal oscillator is 6.655 MHz. Therefore, the output frequency is a function of the carrier oscillator, the VFO, and the crystal oscillator. Since the VFO is at a relatively low frequency and the other frequency-determining oscillators are crystal-controlled, the transmitter is extremely stable.

There are as many choices for the IFs in the dual-conversion SSB transmitter as there are designers. Not all systems place the second local oscillator above the output frequency, and thus some systems have dial reversal and sideband inversion. This does not imply that these systems are in any way inferior. Except for the corrective actions, the performance of systems using a second local oscillator above or below the output frequency are identical.

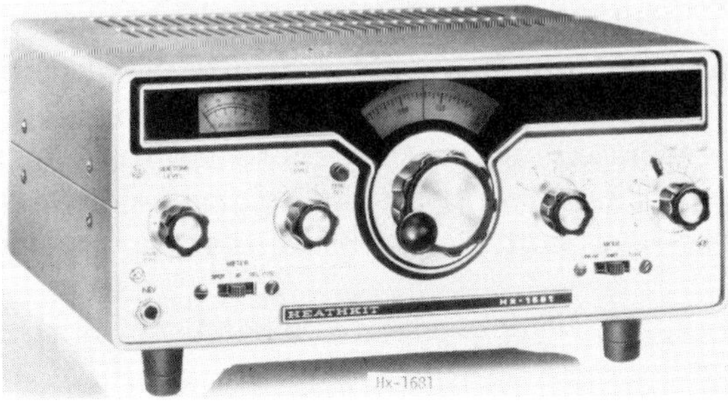

The Heath HX-1681 is a CW-only transmitter that provides about 100-W output. (Courtesy of Heath Company.)

2.3 ACCESSORIES AND AUXILIARY SYSTEMS FOR SINGLE-SIDEBAND TRANSMITTERS

An almost universal accessory for a single-sideband transmitter is CW operation. This mode of operation may be accomplished in two ways. A CW carrier may be generated by the carrier oscillator, which is slightly different from the frequency of the SSB carrier. The CW carrier falls in the center of the sideband filter, while the SSB carrier falls on the upper- or lower-frequency skirt of the filter. The balanced modulator is purposely unbalanced so that the signal from the carrier oscillator passes through the sideband filter. The carrier oscillator signal is then mixed and amplified in the normal fashion. Keying may be accomplished by switching the balanced modulator from the balanced to the unbalanced state, which effectively switches the carrier on and off.

Another method of generating CW in a single-sideband transmitter is to modulate the transmitter with a sine-wave signal applied to the audio input. This modulation, called "single-tone" modulation, produces a single output frequency or CW.

Two potential hazards exist with this method. First, if the audio oscillator used to modulate the balanced modulator has any harmonic distortion whatsoever, there would be a second CW signal, which would be a spurious output. This can effectively be eliminated by using a low-distortion oscillator with a frequency high enough so that the second harmonic is outside the passband of the sideband filter. The second problem is that the dial, which normally reads the carrier frequency, would not indicate the frequency of the CW signal but would be offset by a frequency equal to the audio-oscillator frequency. This can be handled by simply adding a second cursor for the dial for CW use only, or by electronically shifting the VFO frequency slightly for CW operation.

Automatic level control (ALC) is analogous to the AGC system of a communications receiver. To maintain linearity in an SSB transmitter, it is important that none of the amplifier or mixers in the signal path be overloaded. A signal of excessive level would cause a dangerous amount of distortion to be generated and possibly cause harmful interference to other radio services. Therefore, good-quality single-sideband transmitters use an automatic level control to ensure that no circuits are overdriven.

Figure 2–14 shows a simple, single-conversion sideband transmitter with an ALC feedback system.

A sample of the *RF* output voltage is rectified by D1 and filtered by C1. No action takes place until the voltage across C1 becomes greater than V1. The RF level at this point is called the *ALC threshold.*

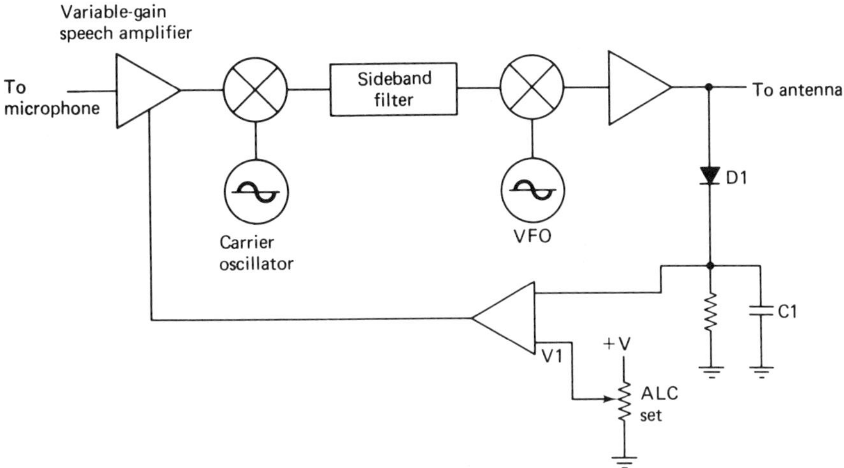

FIGURE 2-14. ALC system using the speech amplifier for control.

Beyond this level of power output, the ALC amplifier causes the gain of the microphone amplifier to be decreased and thus reduces the power output. The output level is effectively prevented from increasing much beyond the ALC threshold level. ALC is considered a protective device, and operating a transmitter with heavy ALC action is not a desired method of obtaining more "talk power."

There are, however, many devices for increasing the average speech power without decreasing intelligibility. The power of normal speech has a very high peak-to-average ratio, which would be reduced in order to increase the "talk power" of an SSB transmitter. One simple way to achieve this is to clip the audio-modulating waveform and very carefully filter the result to eliminate the harmonics. An important advantage of this method is that the speech processing, or clipping, can be done completely outside the transmitter.

A more effective and elegant method of speech processing is called *RF clipping*. In this method, the double-sideband suppressed carrier signal is clipped as it emerges from the balanced modulator, and since the harmonic distortion occurs at twice the carrier frequency, it is very easy to filter. Much heavier clipping may be employed without seriously decreasing intelligibility.

An auxiliary circuit often incorporated into the ALC system is a power cutback for protection against destruction of the power transistors due to excessive heating from high standing waves. A directional coupler or standing-wave-ratio (SWR) bridge is incorporated between the power amplifier and the antenna, and the reflected power signal is used to cut back the drive to the power amplifier when the reflected power exceeds a preset amount. The transmitter will continue to operate but at a reduced power output.

More research and design effort has probably been applied to the variable-frequency oscillator as used for transmitter and receiver frequency control than to any other circuit. Early radio transmitters used quartz crystals for frequency control because the variable-frequency oscillators of the day were not stable enough for reliable communications. The variable-frequency oscillators used in a typical communications receiver were no more stable than those in the transmitter,

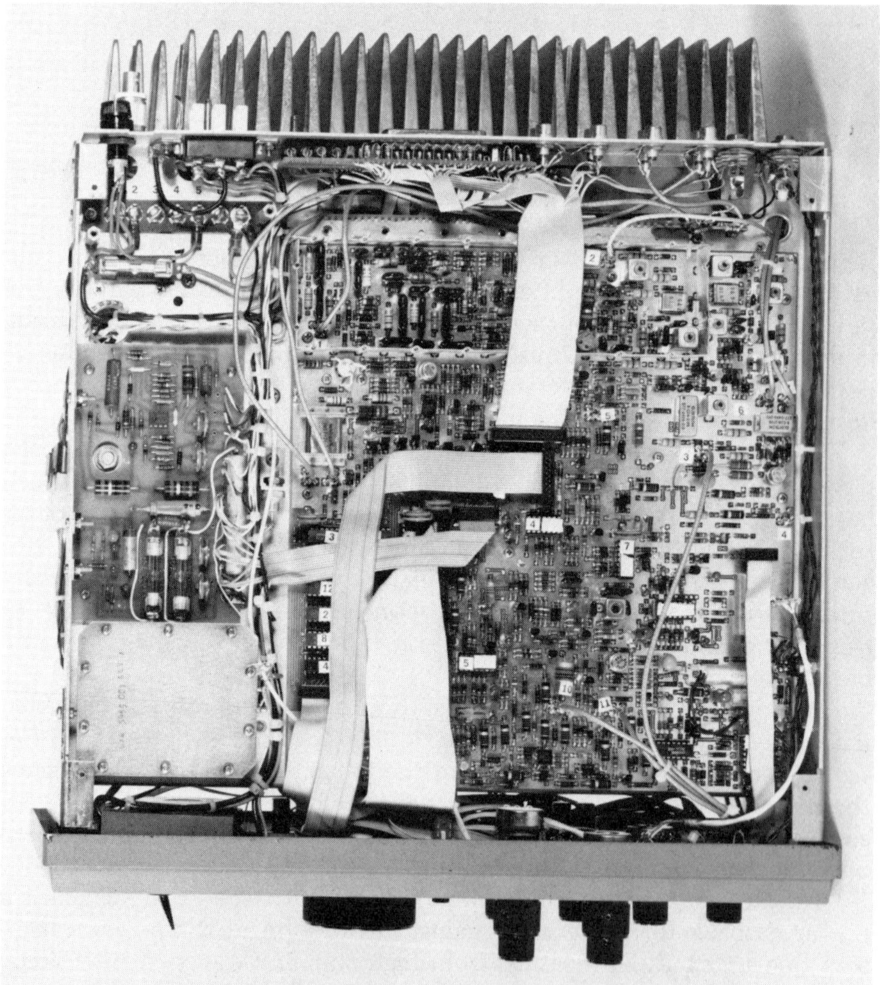

The bottom view of a fully synthesized transceiver, the Collins KWM380. (Courtesy of Collins Group of Rockwell International.)

but at least they could be periodically zero-beat to match the transmitter frequency. For this reason most research into stable accurate VFOs was for use in radio transmitters.

Any basic form of oscillator may be used in the variable frequency oscillator, the only difference being the method by which they are tuned.

The simplest and most popular over the years, although losing ground to more modern designs, is the capacitively tuned oscillator, which is relatively stable, and if constructed with special capacitors, can be very linear. VFO linearity is very important if a gear-driven two-part dial system is to be used. This dial can achieve extremely high resolution for a small investment in panel space. Usually, a calibrated circular dial attached to the tuning shaft makes several revolutions while a second dial reads out the coarser resolution. If the VFO is not linear, inaccuracies will appear after several rotations of the circular dial.

To combat this, special variable capacitors are made with custom-shaped plates to compensate for the normal nonlinearity. Even with very expensive capacitors, it is difficult to mass-produce a good linear oscillator.

A variable oscillator that is very linear is the *permeability-tuned oscillator* (PTO). Instead of tuning the oscillator with a variable capacitor, the frequency of the oscillator is set with a variable inductor. A precision slug is slid in and out of the oscillator coil with a lead-screw type of arrangement. This type of oscillator tuning system is inherently more linear than the capacitor tuning system. It is possible to improve the linearity by winding the inductor in a nonuniform fashion. Even further improvements in linearity are made with electromechanical systems using variable trimmer capacitors.

In some systems it is desirable to have an electronically tuned oscillator. These oscillators, usually varactor-tuned, have a very narrow tuning range because of the high-temperature dependence of the varactor diode. Oscillators of this sort are found in passband tuning systems, variable-frequency BFOs, and so on. For a narrow tuning range, the linearity of the varactor-tuned oscillator can be very good.

Like the communications receiver, the SSB transmitter can be outfitted with a counter for frequency readout. Since the VFO does not operate at the output frequency (except in the case of the simple MOPA CW transmitter), methods are employed similar to those discussed in Chapter 1 for adding or subtracting the IF. Perhaps the only difference between a transmitter readout and one used in a receiver is in the permissible levels of interfering radiated signals. Since there are no very low level amplifiers in the SSB transmitter such as those found in a receiver, the amount of shielding and filtering around the frequency counter may be relaxed somewhat. This does not allow total abandon in the counter design. If the radiated signals are strong enough, they will be heard in the station receiver.

With single-sideband operation, pauses in speech produce a condition of no output power. It is often desirable to switch to a receive mode during this time to approximate a duplex type of operation. This is accomplished using a voice-operated transmit (VOX) mode, where talking into the microphone activates the transmitter. The attack and release times may be adjusted from a fast action, where even short pauses allow the receiver to be enabled, to slow release times of several seconds.

One almost obvious problem is false tripping due to the audio from the receiver itself. This is eliminated by using an anti-VOX circuit. Audio from the receiver is sampled and applied to the VOX circuit in an opposite polarity from the signals from the mircrophone. This self-canceling effect prevents the receiver audio from causing false trips.

In SSB transmitters where the CW mode is generated by feeding an audio signal into the speech amplifier, the VOX circuits are often left intact during CW operation to provide semi-break-in operation.

2.4 TRANSMITTER CIRCUITS

The circuits that comprise a radio transmitter are the same, functionally, as those found in the receiver. Amplifiers, mixers, and oscillators are the basic circuits found in the transmitter and have some obvious and subtle differences from their receiver counterparts. For example, an RF amplifier found in a radio transmitter must handle signals up to perhaps hundreds of watts, whereas the RF amplifier found in a receiver must have a wide dynamic range and usually handles signals in the microvolt range. On the other hand, a mixer used in a radio receiver may be identical to a mixer used in some of the low-level stages of a SSB exciter. Only those transmitter circuits that are significantly different from receiver circuits will be discussed in this chapter.

2.4.1 Speech Amplifiers

The main purpose of a speech amplifier in a SSB transmitter is to raise the low level of the microphone to a level suitable for proper operation of the balanced modulator. The typical output from a dynamic microphone is on the order of 10 to 100 mV, while 200 to 300 mV is required for a balanced modulator. In addition, speech amplifiers often supply some bandwidth reduction and additional output signals for activating a *voice-operated relay* (VOX) system. On

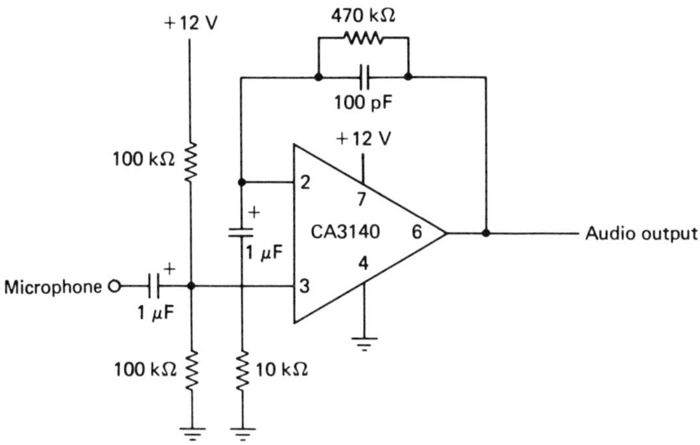

FIGURE 2-15. Simple speech amplifier for a low-cost sideband generator using an operational amplifier.

some ALC circuits, the speech amplifier gain may be controlled automatically for stabilizing the output level.

Figure 2-15 shows a simple speech amplifier suitable for a low-priced SSB transmitter incorporating a rudimentary form of bandwidth reduction. The output of the speech amplifier is capable of driving a high-impedance active balanced modulator from the lowest-level microphone.

The speech amplifier shown in Figure 2-16 uses an AGC integrated circuit in conjunction with a matching speech-amplifier chip. In this way, much of the AGC action of the transmitter is obtained before any

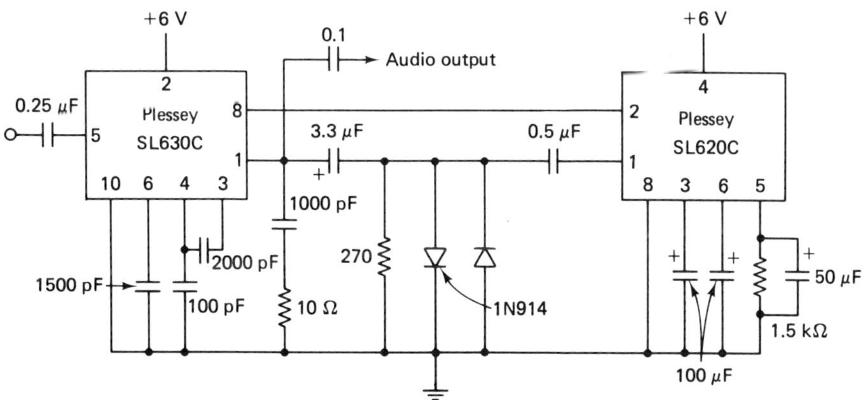

FIGURE 2-16. A speech amplifier using automatic gain control can be used for SSB or any other type of modulation.

of the RF circuits. This reduces the required dynamic range of the remaining gain-controlled circuits.

2.4.2 VOX System

The VOX system, although applicable to any type of voice communications, is applied only to SSB in amateur equipment. Two inputs are required for proper VOX operation: a signal from the speech amplifier and a signal from the receiver. Referring to Figure 2-17, the signal from the station receiver is set in level and amplified. Similarly, the signal from the speech amplifier is adjusted in level and amplified. Both amplified outputs are rectified, except with different polarities. In the example, the VOX signal is rectified into a positive voltage and the anti-VOX signal is rectified into a negative voltage. The two opposite polarity voltages are added together and smoothed with a capacitor. The comparator, U1, will change state only if the VOX voltage is larger than the anti-VOX voltage. Since some of the audio from the speaker will appear in the VOX channel, the VOX and anti-VOX sensitivities must be adjusted so that the receiver does not trip the VOX and that the receiver audio does not prevent the normal VOX from operating.

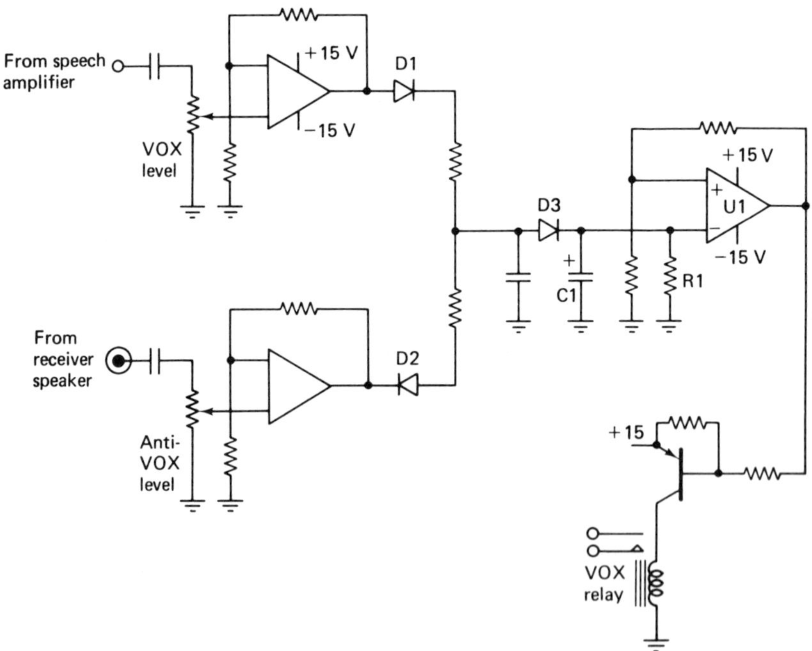

FIGURE 2-17. Complete VOX system, including anti-VOX.

To prevent the VOX from "chattering," that is, going on and off between syllables, a long time constant is provided with R1, C1, and D3. This gives an effect similar to the fast-attack long decay used for AGC systems to prevent AGC pumping. The amount of delay is often an operator preference and can sometimes be set from the front panel.

2.4.3 Balanced Modulators

The double-balanced diode mixer provides one of the simplest-to-use and the best-performing mixers available. A typical transmitter application for the double-balanced mixer is a balanced modulator, as shown in Figure 2–18. The carrier oscillator provides a large signal, typically greater than ½ V across 50 Ω to switch, or "commutate," the diodes. It is important that the "commutating" signal be much larger than the audio level, typically from 3 to 10 times as great; otherwise, serious distortions will occur. To operate properly, the impedances of the driving sources, both the audio and the carrier oscillator, should be low, around 50 Ω. A well-designed double balanced mixer with matched diodes will supply 40 to 60 dB of carrier suppression with a signal loss of about 6 dB. Prepackaged double balanced mixers are available for prices that hardly make it economical to purchase the individual components. These mixers have excellent performance characteristics over a very wide frequency range, but the lower frequency limit often does not include 455 kHz. Double balanced mixers for these low-frequency IFs are often assembled from individual components for that reason, while prepackaged mixers are used for all high-frequency IFs. The mixer transformers must be wound on high-permeability material, and the diodes must be matched silicon junction types or hot carrier

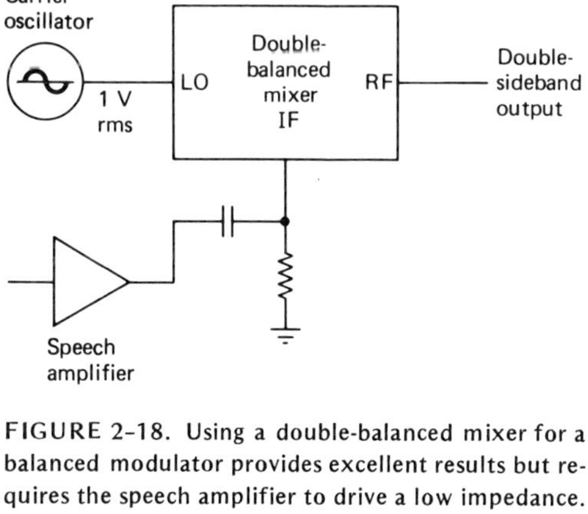

FIGURE 2–18. Using a double-balanced mixer for a balanced modulator provides excellent results but requires the speech amplifier to drive a low impedance.

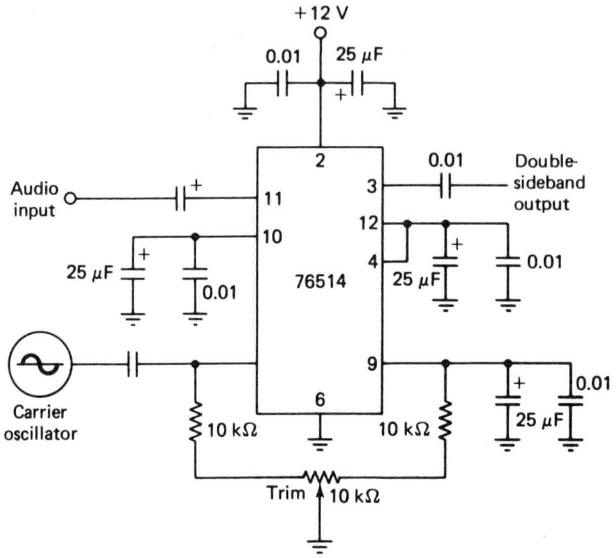

FIGURE 2-19. An active balanced modulator eases the requirements on the speech amplifier and provides some gain.

types. The matching of hot carrier diodes is not necessary for most low-frequency balanced modulator applications.

An active balanced modulator is shown in Figure 2-19. The main advantage of using the active modulator is the available gain. Even though the gain is attractive, the designer must take care not to allow the output signal level to become excessive, because most sideband filters have restrictions on the level of signal that can be safely handled by the filter. To further enhance the carrier suppression, a trim adjustment is included.

2.4.4 IF Amplifiers

The IF amplifier, as found in a SSB transmitter, bears only a slight resemblance to the IF amplifier found in a receiver. The signal levels in a transmitter are much larger than those in the receiver. Therefore, noise figure and gain are seldom a problem, but linearity must be carefully controlled.

Figure 2-20 shows an IF amplifier terminating a mechanical filter. This circuit is similar to an IF amplifier found in a receiver except that degenerative emitter feedback is used to control the gain and improve the linearity. Very little gain is required of IF amplifiers in a SSB transmitter. In a communications receiver, the bulk of the receiver gain

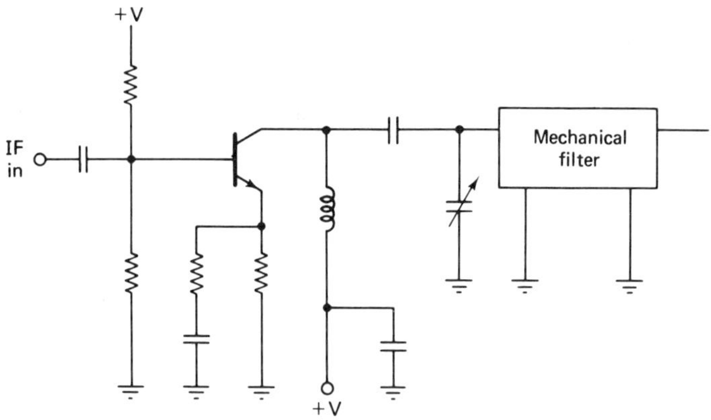

FIGURE 2-20. IF amplifier terminating a mechanical filter.

was obtained in the IF amplifier. An IF frequency must be mixed at least once before it becomes the final operating frequency. Providing large IF gains implies that the resultant signal is very large. Finding a suitable mixer to handle a large IF signal would be difficult and dangerous from the standpoint of radiating images.

2.4.5 Power Amplifiers

The bulk of the SSB transmitter gain is obtained at the final operating frequency by RF amplifiers. All RF power amplifiers used in a SSB transmitter, regardless of power level, must be linear. The best linearity and greatest gain may be had using a class A amplifier, where the current flows through the transistor for the entire RF waveform. This is also the least efficient type of amplifier and is, therefore, used for only the lowest power stages. RF feedback is used on RF amplifier stages to increase the linearity, but it has the undesirable effect of reducing the gain of the amplifier.

For higher-power amplifiers where efficiency is important, class B operation is used. In a class B amplifier, current flows for about half of the RF cycle. The efficiency obtainable from a class B amplifier is about 50 percent, whereas the efficiency of a class A amplifier is about 20 percent. Class C stages, popular for CW transmitters because of their high efficiency, are never used for SSB work because of their very poor linearity.

Figure 2-21 shows a class A tuned amplifier, typical of the type used immediately following a mixer in a SSB transmitter. It is not unusual for amplifiers of this type to use inverse RF feedback as shown, and powers almost never exceed 1 W peak envelope power (PEP). Feedback

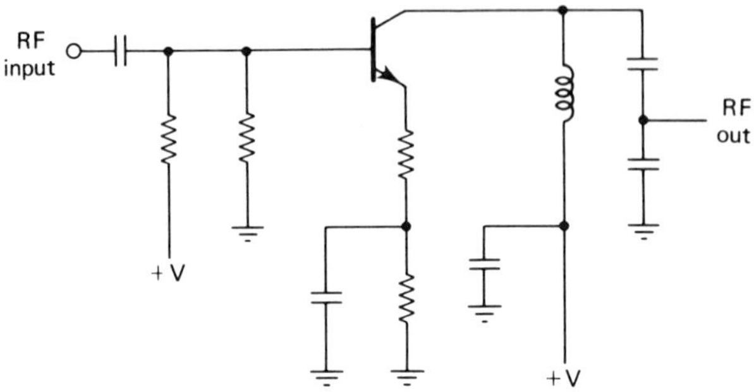

FIGURE 2-21. Class A amplifier with emitter feedback.

in this example is supplied by a partially decoupled emitter circuit. A tuned amplifier almost always follows a mixer stage to ensure complete removal of the undesired mixer product.

There has been a growing trend toward the use of the untuned RF amplifier for the higher-powered transistor stages. There are several good reasons for this move. First, an untuned amplifier, of course, does not have to be tuned. This simplifies the mechanical arrangement and routine operation. Second, the extremely low impedance of a transistor at high power levels makes it almost impossible to construct effective tuned circuits to operate directly with the transistor. Impedance matching is accomplished with special wide-band RF transformers, and any necessary filtering, such as the elimination of harmonic energy, is done after transforming to a higher impedance, such as 50 Ω, and is matched to the transistor with the wide-band transformer.

Figure 2-22 shows a single-ended wide-band RF power amplifier

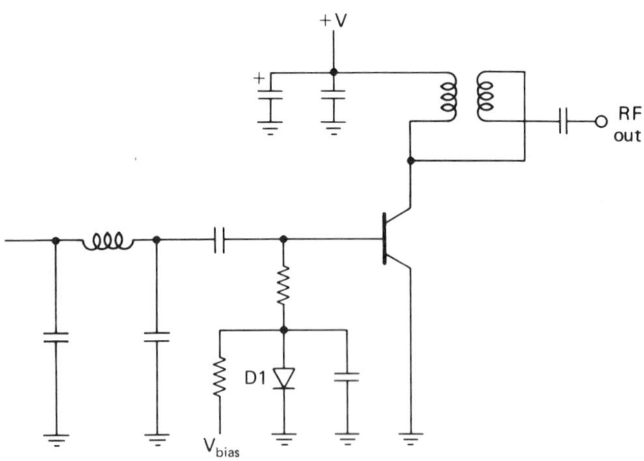

FIGURE 2-22. Class B intermediate-power amplifier with a single-ended output.

with transformer coupling. Circuits of this sort are typically operated in class B.

Generating the bias for a class B amplifier must be done with care. Most amplifiers use a form of thermal feedback to prevent *thermal runaway*. A diode, D1, in Figure 2–22, is placed in thermal contact with the power transistor. This is usually accomplished by mounting D1 on the same heat sink as the RF power transistor. Temperature variations in the RF power transistor cause the voltage drop to change in the diode. As the transistor becomes hotter, the bias voltage decreases and hence the collector current decreases. This reduces the quiescent power dissipated in the transistor and prevents destruction of the power transistor. There are some specialized RF power devices with temperature-sensing diodes inside the transistor package to provide intimate thermal contact with the transistor die inside the package.

Increasing the power-output capability of a transistor amplifier can be accomplished by using a transistor of a higher power capability or by increasing the number of transistors used in the amplifier. One method of increasing the number of transistors is to use two devices in push-pull. Aside from increasing the number of transistors, there are other advantages to the push-pull amplifier. The most significant advantage is the reduction of even-order harmonics, especially the second harmonic, which is the most difficult to filter. A typical push-pull amplifier is shown in Figure 2–23. Dc power supply voltages, both bias

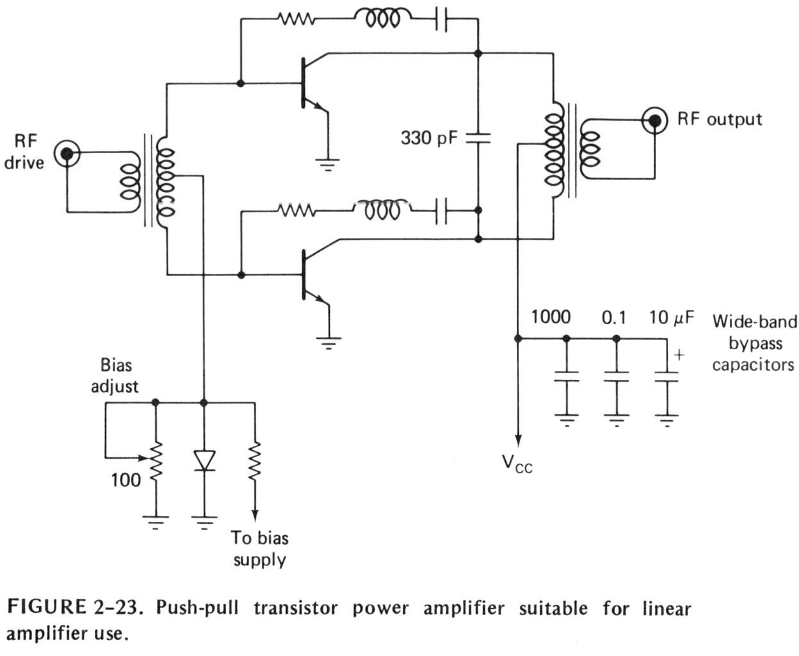

FIGURE 2–23. Push-pull transistor power amplifier suitable for linear amplifier use.

and collector supply, are supplied in parallel to each device through the center taps of the transformers.

The power gain of any transistor decreases with increasing frequency. If there were no frequency compensation in the power amplifier, the gain would be very high at low frequencies and much lower at the high-frequency end of the spectrum. This is undesirable for several reasons. First, the extreme gain at the lower frequencies would increase the possibilities of oscillations at some low frequency, and the drive requirements would be drastically different between the high end of the frequency range and the low end. To counter this effect, negative feedback is provided which tends to stabilize the gain by increasing the feedback at lower frequencies. The series resistor, inductor, and blocking capacitor from each collector and base in Figure 2-22 are the feedback networks.

2.4.6 Power Supplies

The discussion of power supplies could be included in the discussions of any type of amateur equipment. Since power is required by all electronics circuits, power supplies abound. Although very few transmitter or receiver specifications refer directly to the power supply, the ability to meet other specifications can depend on sufficient and reliable power being supplied to all the circuits.

Batteries are the most basic form of power and are used in portable equipment, such as VHF handy-talkys. Two types of batteries are in use: the primary battery, which is not rechargeable, and the rechargeable cell or secondary battery. The *primary cell* is usually an alkaline or zinc-carbon cell that is simply installed in a radio and removed when its energy has been depleted. The *secondary cell* requires proper handling if its full potential is to be realized. Perhaps the most common type of secondary cell in use is the nickel-cadmium cell. These cells have an open-circuit voltage of 1.25 V and possess some very useful characteristics for portable equipment. First, the batteries have a high number of charge/discharge cycles, typically 500, or with careful treatment, more than 1000. Second, the cell voltage does not reduce with the state of charge; that is, a transmitter may provide full power output until the entire charge of the battery has been used. These batteries are available in several forms, the most common being sealed cells that from the outside are similar in appearance to ordinary flashlight batteries.

Maximum battery life can be realized only if the batteries are charged and discharged properly. Nickel-cadmium batteries should be charged at about one-tenth of their rated discharge rate. The capacity of a battery is rated as the amount of current the battery can continuously supply for 1 hour. For example, a typical nickel-cadmium cell about the size of an AA flashlight cell has a capacity of about 0.4

ampere-hour. This means that if the cell were subjected to a load of 0.4 ampere (A), it could supply this current for 1 hour. If half of that current were required, the cell would last roughly 2 hours, one-fourth of the current and the cell would last about 4 hours, and so on. Usually, the ampere-hour capacity will increase slightly as the current drain is decreased. When charging a 0.4-A-hour battery, the charging current should be about 40 mA or one-tenth of the 1-hour discharge current. Realizing the inefficiencies of the battery, the total charge time will be about 16 hours.

For maximum life, the batteries should not be discharged beyond a point where the battery voltage drops more than about 10 percent below the fully charged value. When nickel–cadmium cells reach discharge, the cell voltage drops suddenly. Some battery-operated equipment is supplied with a battery voltmeter which will clearly indicate a depleted battery. In many cases there will not be a meter, and obvious changes in equipment performance will have to be used as indicators. A sudden drop in transmitter power, an FM receiver that will not squelch, and low-frequency oscillations of an audio amplifier are typical indications of a depleted battery. Once it has been determined that the battery has been depleted, it should be charged immediately.

There is an effect unique to nickel–cadmium batteries called the *memory effect*. If these batteries are charged and only a small amount of charge is removed, and they are recharged repeatedly, the ampere-hour capacity of the battery will be temporarily reduced. This effect can be reversed completely by discharging the battery completely before recharging, and doing this several times, before returning the battery to normal use.

Most amateur communications equipment will not operate directly from batteries but will be supplied dc power from a power supply oper-

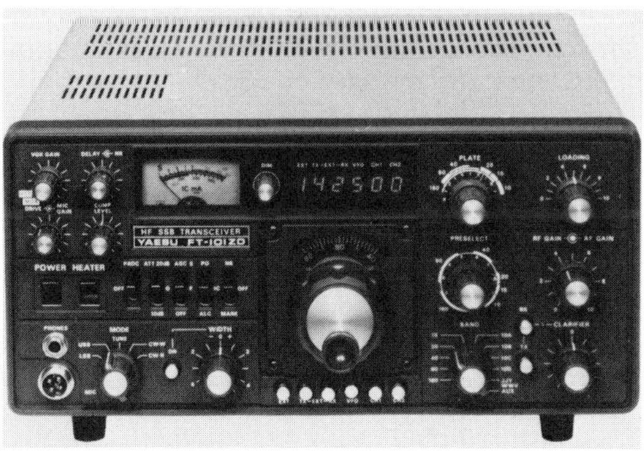

The Yaesu FT-101ZD is a popular transceiver featuring digital readout. (Courtesy of Yaesu Electronics Corp.)

ating from alternating current. Some circuits require that the power be conditioned and stabilized, whereas other circuits will tolerate a certain amount of voltage fluctuation and ripple. Some stages in a transmitter or receiver, such as oscillators, AGC systems, and audio amplifiers, are very sensitive to voltage variation. Other stages, primarily power amplifiers in both receivers and transmitters, do not require regulated voltages, but the latest trends are toward completely regulated power supply voltages for every stage in a transceiver. In many cases, the addition of voltage regulation requires nothing more than a simple three-terminal device, but for other applications, the three-terminal device will not provide sufficient current or the measure of performance required, and a more sophisticated regulator must be used.

Figure 2-24 shows the block diagram of a typical voltage regulator. This block could represent the inside of a three-terminal device or may be implemented with discrete transistors and integrated circuits into a high-current or low-noise specialized circuit. The output voltage is compared to a stable voltage reference. The output of the error amplifier, which is proportional to the difference between the two voltages, is used to increase the current to the load until the voltages are exactly the same. The frequency response of the amplifier and the pass transistor are usually made fast enough to remove most of the ripple from the power supply. A full-wave rectifier and filter would be used to supply the voltage regulator. Full-wave rectifiers have the advantage of a higher ripple frequency and lower current in the individual diodes.

In mobile applications, it is often necessary to supply high voltage for a vacuum-tube amplifier from the 12-V vehicle supply. A transformer is used, but the ac that is required for transformer operation is supplied by transistors which are driven either from an external driving circuit or from a feedback winding on the transformer. Figure 2-25 shows a simplified schematic of a high-voltage power supply that also incorporates voltage regulation. By modulating the pulse width of the pulses supplied to the transformer, the driver circuits can be made not

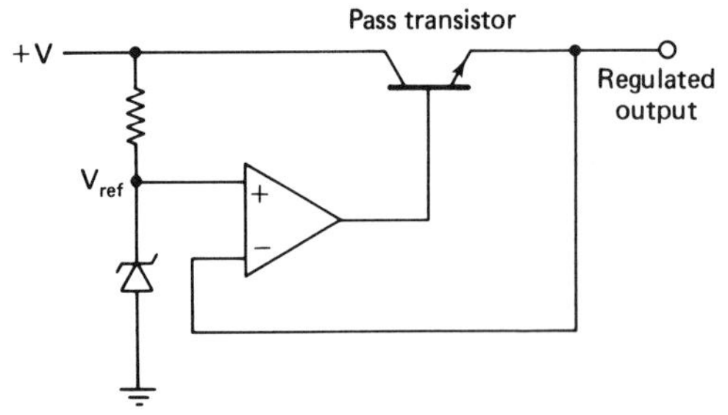

FIGURE 2-24. Simplified schematic of a voltage regulator often called a "linear regulator."

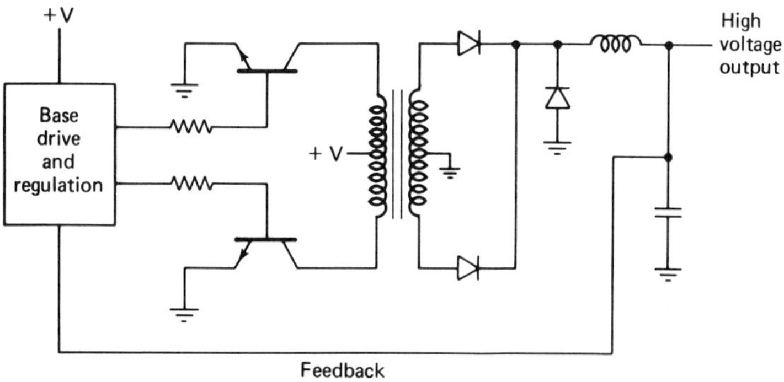

FIGURE 2-25. Power supply using a "switching regulator."

only to provide the drive for the transistors but also to regulate the output voltage. Switching power supplies of this sort can be designed for extreme high efficiencies and transmitter power levels, up to the maximum legal limit. Because of the high switching frequency involved, typically over 20 kHz, the size of a full-kilowatt power supply can be very small. The transformers, filter inductors, and capacitors are much smaller than those encountered in 60-Hz power supplies. To take advantage of this small size, power supplies for use with ac power use high-frequency switching techniques and simply rectify the ac line voltage to supply the dc power.

Special care must be taken by the designer of a switching power supply to use the right components. Semiconductor-device parameters often neglected by the 60-Hz power supply designer can spell the difference between an efficient power supply or a destroyed power supply. Reverse recovery time of diodes and switching times of transistors are two very important parameters that must be known. Another very important design requirement for high-frequency switching power supplies is adequate shielding. The harmonics of the basic switching frequency can be a strong source of interference in a communications receiver. Their elimination requires shielded compartments and special feedthrough filters for all leads entering the shielded area. As the cost of energy becomes greater, the use of efficient power supplies for high-power equipment will increase.

2.5 SINGLE-SIDEBAND TRANSCEIVERS

Originally conceived as a method of cost reduction, the modern SSB transceiver offers ease of operation, small size, and uncompromised performance. Throughout the discussions of superheterodyne receivers and heterodyning transmitters, reference was made to the similarities between this type of transmitter and receiver. The transceiver uses this commonalty to share circuits and to effectively combine as much of the transmitter and receiver as is practicable.

A very simple one-band single-conversion transceiver is shown in Figure 2-26. In this transceiver, the carrier oscillator, local oscillator, IF filter, and IF amplifier are shared between transmit and receive. Not all possible sharing has been used; only those circuits that have large cost benefits or operating benefits are shared. In this example, the product detector and the balanced modulator could be one circuit, while the transmitter and receiver mixers could be shared. Often, the amount of extra circuitry required to perform the switching between transmit and receive is more than what is saved by the circuit sharing and is hardly worth the effort.

Oscillators and filters must be shared to gain any advantages from a transceiver. The operating frequency of a receiver is determined by the frequency of the local oscillator, the beat-frequency oscillator, and, to some extent, the frequency of the IF filter. The frequency of a heterodyning transmitter is, similarly, a function of all its oscillators. Using the same internal oscillators for both transmitter and receiver assures that both units will be on exactly the same frequency and sideband. This is especially useful for SSB operation, where spotting the transmitter is a difficult task.

The requirements of a sideband filter for transmitter or receiver use are exactly the same and allow one of the more expensive components of a transmitter or receiver to be shared without any operating compromise. Usually, the inputs and outputs must be switched between transmit and receive, and this can be accomplished with relays or diode switching. Great care must be taken to preserve the ultimate rejection of the receiver by not allowing any signal energy to leak around the switches during the receive mode.

Oscillators are very easily switched between transmit and receive with relays or diodes. It is important not to cause frequency changes due to different loading between the transmit and receive modes. This would cause the transmit frequency to differ from the receive frequency and destroy one of the more valuable characteristics of the transceiver. This is accomplished by assuring that sufficient buffer amplifiers are ahead of the switching or that the transmit and receive loads are identical.

In a deluxe or sophisticated transceiver, where there are a lot of aids and operating conveniences for the receiver, such as passband tuning and notch filters, these circuits must be disabled during periods of transmit. If there are several different IF filters available for the receiver, these must be switched out of the circuit during transmit and the normal SSB filter reinstated. More complex transceivers must use

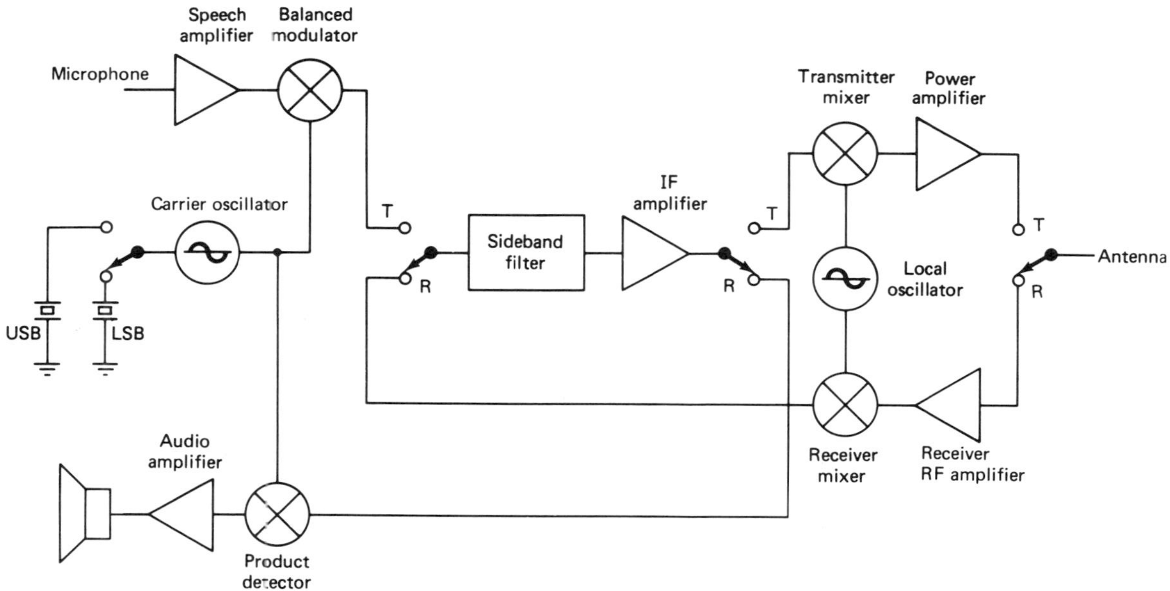

FIGURE 2-26. Single-conversion single-sideband transceiver.

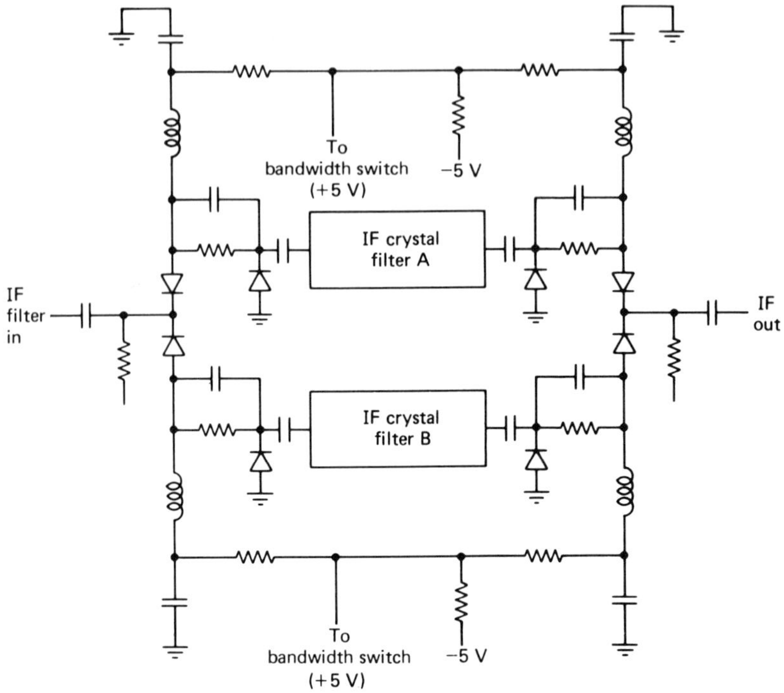

FIGURE 2-27. Solid-state filter switching is required for the more versatile transceivers.

diode switching to accommodate the large number of switching tasks required. Although normal signal diodes may be used for RF switching, it is best to use PIN diodes which are manufactured especially for RF switching. An IF filter switching circuit using PIN diodes is outlined in Figure 2-27.

When the PIN diode is forward-biased, it has a very low impedance to RF current, often less than 1 Ω. In the reverse-bias condition, the resistance is very high and the diode capacitance is very low, often less than 1 pF. However, PIN diodes are not as effective at low frequencies as they are at higher frequencies, especially at VHF. Their use at 455 kHz would not be desirable, and switching done at this IF is best left to relays. In the figure, the filter that is selected has forward-biased diodes at both the input and the output, while the nonselected filters have reverse-biased diodes at both ends.

One of the advantages of the transceiver can also be a disadvantage for some situations. The fact that the transmitter and receiver are always on the same frequency can be a disadvantage if the distant station does not have a transceiver or there are problems with that station's transceiver so that it is not receiving on exactly the same frequency that it is transmitting. An independent control on the receiver frequency is required to correct for the separation between transmit and receive frequencies. This is accomplished by tuning the local oscillator only during the receive mode and returning the local oscillator to the normal frequency during transmit. This is usually accomplished by a varactor diode in the local oscillator circuit, as shown in Figure 2-28. In this circuit, the varactor diode is biased to a nominal voltage during transmit and when the receiver incremental tune (RIT) is off. When the RIT control is being used, the varactor changes the local oscillator several kilohertz only during receive periods. The actual received frequency will not be displayed on the dial unless a digital readout is being used, so it is desirable not to allow the RIT to have a very large tuning range.

It can be seen that a number of functions must be altered between transmit and receive in a large SSB transceiver. It is important that as many circuits as possible be electronically switched, rather than relying on relays for the task. Older transceivers used multiple-pole relays, which were one of the high-failure items in transceivers. The modern transceiver uses diodes, transistors, and analog gates to accomplish the task reliably.

Even with RIT, the frequency agility of the transceiver may not satisfy some operators. Therefore, a remote VFO is available as an accessory to most transceivers; this may be used for controlling the receiver alone, the transmitter alone, or the transceiver. This device allows the transmitter to be separated from the receiver frequency by any amount of frequency. It also allows a frequency to be set by the remote VFO and selected by a switch for monitoring a second frequency with the receiver. The remote VFO is usually nothing more

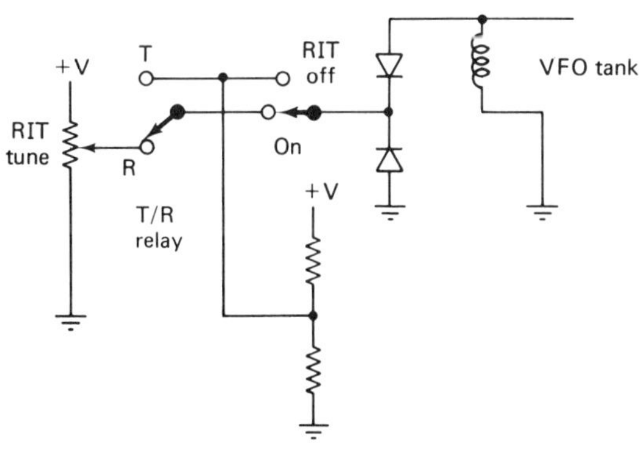

FIGURE 2-28. Receiver incremental tune (RIT) using a pair of varactor diodes.

than a second VFO identical to the one in the transceiver. In the case of a digital readout, the expense of a second digital readout is not carried and the external VFO has a more coarse analog dial, while the actual frequency of the external VFO is read off the internal digital readout.

With the exception of those circuits required for circuit sharing, most of the transmit and receive sections of the transceiver are very conventional. One exception to this is the power supply, which is required to provide not only for a receiver but a transmitter as well. The combined voltage requirements can cause the power supply to be rather complicated and large.

High current is required for the power amplifier, while lower current but regulated voltage is required for the receiver and the driver stages. A problem not unique to the transceiver but common to all higher-powered solid-state transmitters is the problem of switching the heavy current for the power amplifier from the front panel. Transceivers having power inputs in excess of 250 W are not uncommon and require more than 20 A from a 12-V supply. Switches capable of handling this amount of current are very large and do not fit well on portable transceivers. One solution is to use a relay, but this wastes the power dissipated in the relay coil. Another solution is to allow the power supply to be connected to the power amplifier at all times and to switch the lower-power stages. The current required by the amplifier is very low when there is no drive. There exists one danger, however: the power supply voltage is always present inside the amplifier compartment, and any failure of a part or accident while troubleshooting can be dangerous.

2.6 FREQUENCY SYNTHESIZERS

There has been an increasing number of truly synthesized transceivers appearing on the market for amateur use. The truly synthesized transceiver is to be distinguished from a phase-locked local oscillator, where the frequency control is accomplished with a conventional VFO, which is phase-locked to a local oscillator at some other frequency. The truly synthesized transceiver selects frequencies digitally and can operate on discrete frequencies called *channels.* For high-frequency SSB operation, the channel spacing must be less than 100 Hz and greater than 10 Hz. In a synthesized transceiver, the frequency readout is, of necessity, digital, and frequency drift is essentially nonexistent.

For SSB operation on the high-frequency bands, the input frequency is programmed using an optical shaft encoder that increments or decrements a digital counter which programs the synthesizer. This gives a feel similar to the conventional VFO but with all the characteristics of a frequency synthesizer. In addition, it is possible to program frequencies via a computer for the purpose of memorizing frequencies, automatically scanning the band and other chores. Also, frequencies may be entered via a keyboard, rotary switches, thumbwheel switches, and so on.

The thing that characterizes a synthesized rig from a conventional transceiver is the large amount of circuitry required to implement the synthesizer. The synthesizer used for a SSB transceiver is not a simple system, because of the low noise and high performance required for SSB operation. In addition, the lockup time and switching characteristics must be well controlled because in normal operation the synthesizer will be continuously tuned rather than as with the VHF synthesizer, which is set to one channel and left alone. This usually gives rise to multiple loops with VCOs operating at high frequencies so that the lockup time and overshoot can be controlled.

There has been no standard synthesizer scheme developed to date so that two synthesizers will be presented as representative of a typical high-frequency SSB transceiver.

The simpler of the two synthesizers is used in the Swan Astro 150 transceiver and is representative of a medium-priced synthesized transceiver. The second synthesizer, used in the Collins KWM-380 transceiver, represents a high-priced unit. The Astro 150 synthesizer uses 100-Hz steps and is used for amateur-band coverage only, whereas the Collins synthesizer uses 10-Hz steps and is used for continuous coverage from 500 kHz to 30 mHz. The Astro uses a clarifier control, whereas the 10-Hz steps in the Collins are fine enough to eliminate this control.

Figure 2–29 shows a block diagram of the synthesizer used in the Swan Astro 150 SSB transceiver. This is a single-conversion unit using a 9.000-MHz IF and employs a unique tuning system that uses a microcomputer for frequency control. A system of tuning that allows the rate of change of the frequency to be controlled from the front panel via a potentiometer, called variable-rate scanning (VRS) by Swan, is used to allow easy and controlled tuning in increments of 100 Hz. Two phase-locked loops are used in this scheme as shown in Figure 2–29. One loop is controlled by the band switch; the other loop is controlled by the microcomputer from information supplied by the front-panel controls and the variable-rate tuning program in the computer. The numbers displayed on the front-panel readout are determined by the microcomputer from information from the band switch and the programmable counters in the tuning phase-locked loop. The band phase-locked loop supplies frequencies in multiples of 500 kHz and is controlled from the band switch. The actual local oscillator is the sum

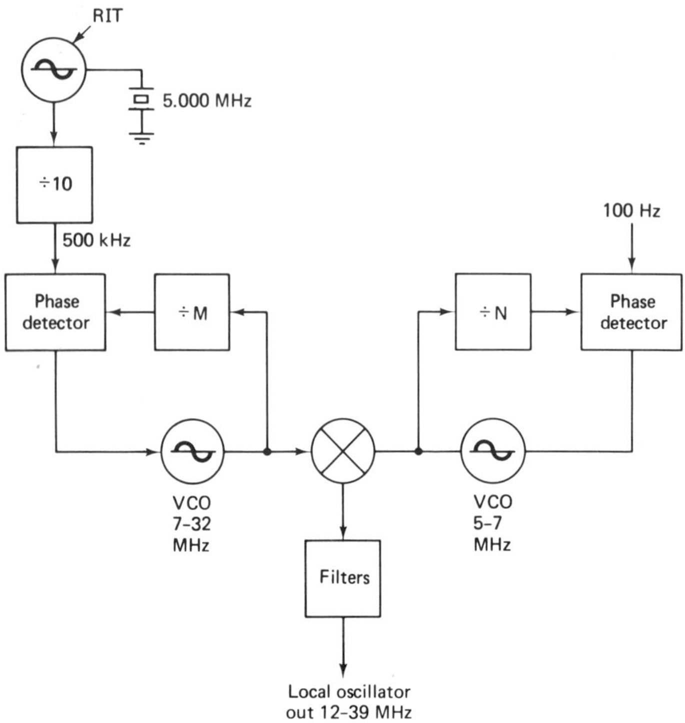

FIGURE 2-29. A 100-Hz-per-step frequency synthesizer with RIT.

of this frequency and another phase-locked loop called the tuning phase-locked loop. The output frequency from the band loop is 7.000 MHz for 80 m, 11 MHz for 40 m, 18 MHz for 20 m, 25 MHz for 15 m, and 32 MHz for the 10-m band. This frequency is mixed with the tuning phase-locked loop, which covers the range of 5 to 7 MHz, and allows up to 2-MHz coverage on each band. The sum of the band loop and the tuning loop is used for the local oscillator output. This signal must be very carefully filtered if the receiver is not to suffer from serious spurious responses.

Receiver incremental tune, which is often difficult to apply to a phase-locked-loop tuning system, is implemented with the reference oscillator of the band loop.

The synthesizer shown in block diagram form in Figure 2–30 is a high-performance synthesizer used in a high-priced transceiver manufactured by the Collins Division of Rockwell International. This synthesizer provides a 39.145- to 69.14499-MHz local oscillator for an up-converting transceiver, covering 0.5 to 29.99999 MHz in 10-Hz steps. The frequency programming information is generated by a microprocessor from an optical shaft encoder.

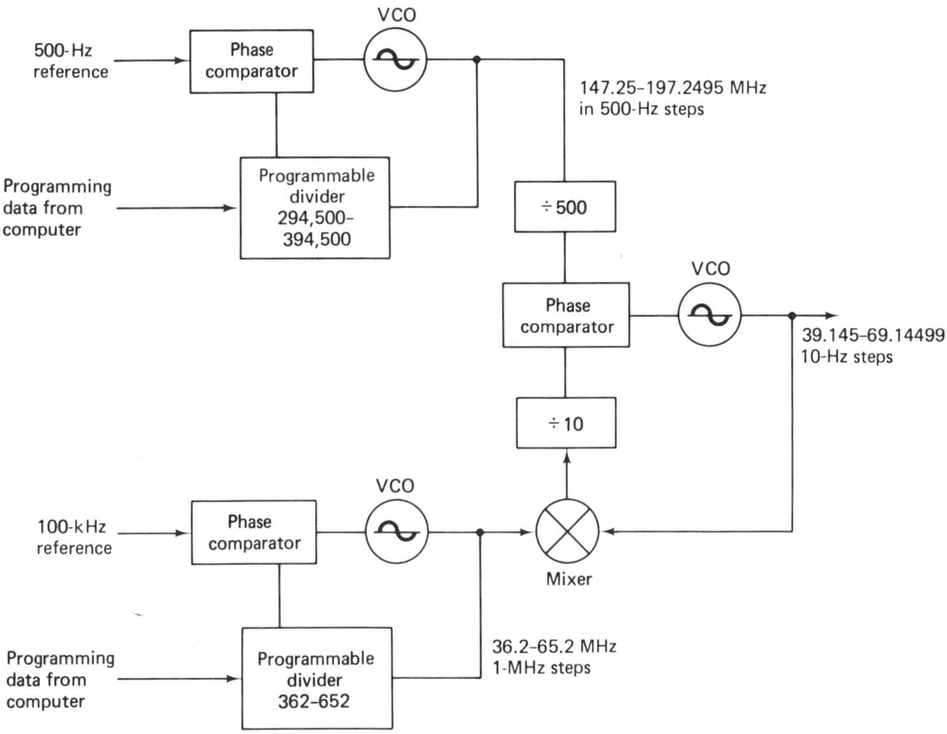

FIGURE 2-30. Block diagram of the synthesizer used in the Collins KWM 380.

Three phase-locked loops are used to generate the entire frequency coverage. Two of the loops are programmed by the computer, while the third loop is controlled by the other two loops. The coarse-tune loop generates frequencies from 36.2 to 65.2 MHz in 1-MHz steps. A 100-kHz reference is used in conjunction with a programmable divider that divides by 362, 372, 382, and so on up to 652. A second loop, the fine-tune loop, generates frequencies from 147.25 to 197.2495 MHz in 500-Hz steps. This is a conventional phase-locked loop using a 500-Hz reference and division from 294,500 to 394,499. As with the coarse-tune loop, the programmable divider is controlled by the computer. The output of this loop is divided by 500, which produces an output frequency from 294.5 to 394.499 kHz in 1-Hz steps.

In the main loop, the VCO is mixed with the coarse-tune VCO, divided by 10, and compared to the fine-tune loop. Therefore, the difference between the main VCO and the coarse-tune VCO divided by 10 equals the fine-tune VCO divided by 500.

It may appear to be a lot of phase-locked loops and rather high frequencies, especially in the fine-tune loop, for generating a local oscillator for a transceiver that operates below 30 MHz. The choice of frequencies was to ensure minimum spurious outputs and to reduce the lockup time for the synthesizer. The fine-tune loop is operated at a very high frequency and then divided down by a factor of 500 so that the reference frequency is 500 Hz and the loop can have a relatively short lockup time. Even with the 500-Hz reference, the fine-tune VCO must be very carefully designed for low microphonics and phase noise. The coarse-tune loop uses a relatively high frequency reference (100 kHz), and thus the VCO is not as critical and the lock-up time is very fast. In the main loop, the phase detector operates from 294 to 394 kHz, which is sufficiently high to minimize lock time for the main VCO.

The microcomputer is not a novelty or a marketing gimmick in a transceiver of this type. It would be nearly impossible to provide a synthesizer of this complexity without the computer to calculate the proper data inputs for the synthesizer.

2.7 HIGH-POWER TUBE-TYPE LINEAR AMPLIFIERS

Most solid-state amplifiers supply a power output slightly more than 100 W CW or PEP on SSB, which represents an input power of about 200 W. Solid-state amplifiers of this power level are seldom class C and are usually linear for SSB operation. Operating higher power levels is usually accomplished by the addition of an external add-on tube amplifier. These amplifiers usually span the power range 1000 to 2000 W PEP power input and are designed for linear operation only. Linear amplifiers may be used for all modes of operation, provided that restrictions concerning duty cycle are observed for some of the modes requiring long periods of power output, such as RTTY and slow-scan television.

The power ratings of amplifiers can be confusing and ambiguous. First, there is the legal limit, which must ultimately be the maximum permissible operating power. This power is defined as 1000 W average input power, which translates into 2000 W PEP, input for SSB, 1000 W key down for CW, and 1000 W carrier for slow-scan television and RTTY. Although the normal rating of peak power to average power for SSB is 2:1, certain speech-processing equipment can alter this ratio to somewhat less than 2. Regardless of what the PEP power is, the average input power must not exceed 1000 W. Amplifiers capable of operating at the legal limit of 2000 W input PEP for SSB service will not, necessarily, operate at the maximum legal power in CW or RTTY service. This is due to the greater duty-cycle requirements of those modes of operation, which tends to increase heating in the amplifier components.

Most external power amplifiers operate in the linear class B mode. There are several good reasons for operating an amplifier in the linear mode even for services that do not require linear amplifiers such as CW. First, for SSB operation, linear amplifiers are required. But for CW operation, the keying characteristics are preserved, which obviates the need for keying circuits in the amplifiers and the necessary interfacing between the driver and the amplifiers. Linear amplifiers with no tuning on the input side are easy to construct, stable, and make good projects for home construction.

There is a major drawback to using a linear amplifier for modes that do not require a linear amplifier: the relatively low efficiency of these amplifiers. A class C amplifier, if carefully constructed, can provide an efficiency of up to 80 percent, with 75 percent being typical. On the other hand, class B amplifiers can supply a maximum efficiency of 65 percent and, typically, 55 to 60 percent.

The use of a grounded-grid amplifier can increase the overall efficiency of the power amplifier because some of the drive power is transferred to the output. For example, if a typical legal-limit amplifier operating at a carrier level of 1000 W requires 100 W of drive, the total output power is about 700 W, which is comparable to the output of a good class C amplifier. The grounded-grid amplifier is a very stable circuit if a good vacuum tube is used. The major drawback to the grounded-grid circuit is the relatively high drive power required for operation.

Class C amplifiers have specific applications for FM, RTTY, and slow-scan television. Because of the requirement for a continuous carrier in these modes, efficiency is a very important consideration in power amplifier designs. A linear amplifier, aside from providing a lower power output, requires that more heat be removed from the amplifier. Since there is no advantage to using a linear amplifier for FM, slow scan, or RTTY, a class C amplifier is the most efficient method of generating high power.

Regardless of the amplifier class of operation or power level, a power supply of sufficient capacity is required for spurious-free operation. Most of the high-powered linear amplifiers require between 2000 and 4000 V with peak and average currents based on power level. Almost any type of rectifier or filter may be used for a high-power amplifer. However, some special considerations must be taken for linear amplifiers. During the voice peak, the power supply voltage must not decrease so low, to allow for clipping. This will cause serious distortions and splatter. It is important that the filter capacitor be large enough to support the power supply voltage during large voice peaks. Class C amplifiers do not require as much capacitance as linear amplifiers, because there is no danger in clipping peaks. A choke input filter may be used for the class C amplifier where a capacitor input filter is usually found in linear amplifier power supplies. Full-wave rectification is always used to get the most from the power supply.

The physical size of the high-power amplifier power supply can depend on the choice of components. Power supply transformers vary in size according to the type of steel used in the core. Modern transformers using high-grade steel can be small enough for a desktop 1-kW amplifier with a self-contained power supply. The penalty for a small transformer is an increase in operating temperature, so much so that forced-air cooling is often required, not only for the vacuum tube but for the power transformer. The size of filter capacitors can also affect the size of the amplifier. Oil-filled capacitors are often used in high-voltage power supplies; however, smaller and less expensive electrolytic capacitors are replacing the older oil-type capacitors. Electrolytic capacitors are seldom made for voltages higher than about 500 V; therefore, it is necessary to place several capacitors in series to obtain the desired working voltages. Placing capacitors in series not only increases the voltage rating of the resultant capacitor, but decreases the capacity value. For example, if five 500-V capacitors were placed in series to obtain a 2500-V capacitor, the value of the resultant 2500-V capacitor would be only one-fifth of the individual capacitance values. Therefore, to obtain a certain value at 2500 V requires that each individual capacitor be five times that value. Even with these large-value capacitors, the end result can be less volume than an equivalent oil-filled capacitor. It is necessary to place resistors across each capacitor to ensure an equal distribution of voltage across the capacitor bank.

A similar method of placing low-voltage components in series for high-voltage operation is used with silicon-diode rectifiers. Silicon diodes with working voltages higher than 1000 V are not generally available, and amateur power amplifiers use series diodes to increase the working voltage of the diode string. As in the case of the capacitors in series, equalizing resistors are used to distribute the voltage. Capacitors are also placed across the diodes to equalize any transients that may appear on the transformer and to compensate for differences in reverse recovery time.

Figure 2–31 shows a typical high-voltage power supply suitable for amateur use, which makes use of series-connected electrolytic capacitors as well as series-connected diodes. Safety interlocks and a time-delay circuit prevent the high voltage from being applied before the vacuum tube has reached operating temperature. The plate voltmeter is connected to the bottom of the bleeder resistor string while the plate current meter is placed in the secondary center tap of the power transformer. A diode is placed across the current meter to prevent any high

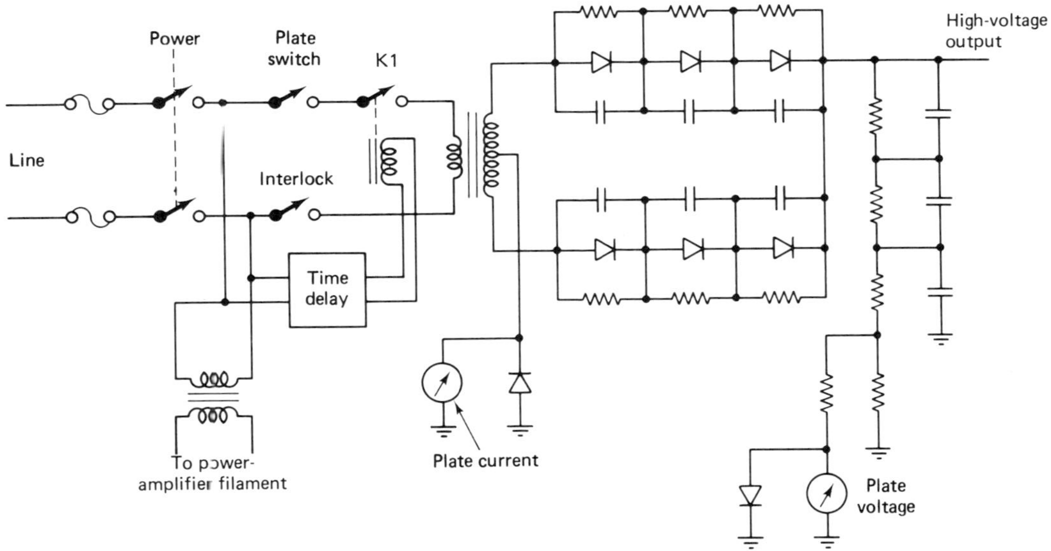

FIGURE 2-31. High-power amplifier power supply, showing protective circuits.

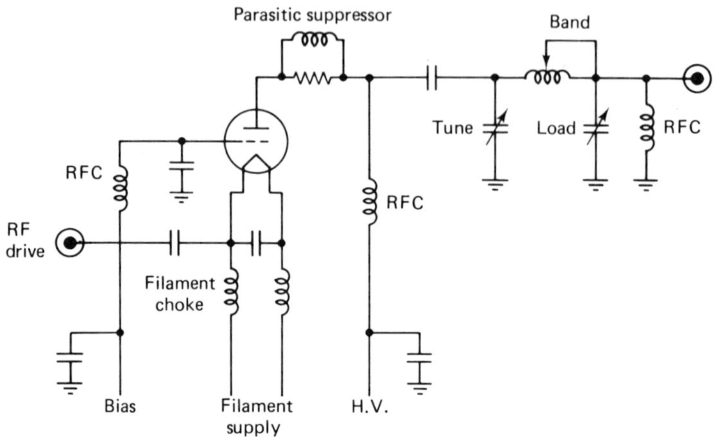

FIGURE 2-32.

voltage from appearing across the meter terminals if the meter should fail and across the voltmeter if the bottom bleeder resistor should fail.

Figure 2-32 shows a power amplifier using a triode in a grounded grid, which is a very popular amplifier for home construction because of its stability and ease of construction.

In this circuit, the vacuum tube has a directly heated cathode which requires the use of a special choke for the cathode circuit. The current for the heater must pass through the cathode choke, which can be current enough to saturate many ferrite materials. To prevent this, the chokes are bifilar-wound such that the current to the heater passes in opposing directions and does not cause the ferrite-core material to saturate. The RF choke at the antenna terminals is not required by the circuit but is placed at that point to prevent dangerous voltages from appearing at the antenna terminals should the plate blocking capacitor fail. The parallel combination of a resistor and an inductor at the plate of the vacuum tube is a parasitic suppressor and is required to prevent oscillations at very high frequencies. This circuit is especially required if the tube is of a high-gain type, such as a ceramic pentode or tetrode.

Tube manufacturers have developed special *zero-bias tubes* for linear amplifiers. This type of tube could be used in this circuit by simply connecting the grid directly to ground. Obtaining a good ground connection for the grid of the tube is very important and the use of a zero-bias tube eases the job of making a good RF ground for the grid.

Tetrodes and pentodes may be used in grounded-grid amplifiers by connecting the screen and the control grid together, or both to ground. Many tubes will give excellent service in this circuit connected as high-gain triodes. However, some vacuum tubes may not be safely connected as triodes; these are usually high-frequency ceramic tetrodes. It is possible to operate these tubes as grounded-grid amplifiers by using a screen supply but placing the screen at RF ground. This is not a true triode connection but will allow the tetrode to be used in a grounded-grid amplifier.

Figure 2–33 shows a grounded cathode linear amplifier with a Pi-L output. This circuit may be used for either class C or linear operation by simply changing the grid bias voltage. This is not possible with grounded-grid amplifiers. Since the drive energy is not transferred to the antenna in this circuit, the overall efficiency is not as great as that of the grounded-grid circuit. This circuit, which has a high-gain ceramic tetrode, can have considerable gain and must be constructed very carefully if the amplifier is not to become a 1000-W oscillator. It is not unusual for less than 5 W of drive to provide full power output in this amplifier. For maximum gain, the 50-Ω input is matched to a much higher impedance in the grid circuit. However, this requires neutralization to ensure stable operation, which effectively lowers the grid circuit impedance. The technique shown in Figure 2–33 reduces the gain of the amplifier but assures stability for all operating conditions. In this circuit, the grid is untuned, as in the grounded-grid amplifier, with a low impedance, typically 50 Ω. This low impedance makes it almost impossible to obtain sufficient feedback to sustain oscillation but requires that several watts be consumed in the resistor. It is very important that the resistor be of a noninductive type.

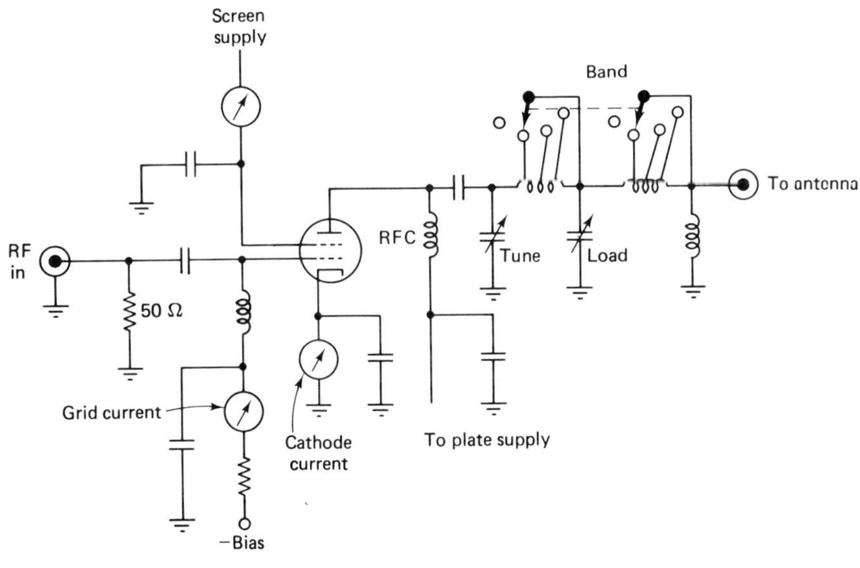

FIGURE 2-33. Grounded cathode linear amplifier with a π-L output.

It is the intent of this book to remove some of the fear and ignorance regarding electronic circuits and systems with the hope that many amateurs will attempt to understand and appreciate their equipment by building transmitters and receivers and repairing equipment purchased commercially. There is one area of electronics in which fear of the unknown should *never* be removed, and that is the area of high-power and high-voltage circuits. First, and it should come as no surprise to anyone, the high voltages present in a 1-kW amplifier will kill; not possibly, maybe, or perhaps, but assuredly. When troubleshooting a high-power amplifier, do not trust your life to the bleeder resistor in the power supply. Before entering the amplifier, remove all power supply plugs from the ac supply and short the high-voltage points to ground, using a grounding rod. Not only is there high voltage within the high-power amplifier, there are points of high temperature. Serious burns can result from brushing up against a vacuum tube that was recently dissipating several hundred watts. When constructing a high-power amplifier in the home shop, never omit any safety-related component. Be sure to use only the components specified and never leave out a part whose purpose is not understood. Many components *appear* to have no purpose in an amplifier; they are present only for protection of components and personnel.

There are other safety hazards not due to the high voltage as such, but due to the high power available from the power supply. The amount of energy stored in the filter capacitors is very great, and long before the line fuse could blow due to a short circuit in the amplifier, the energy stored in the filter capacitors could have spread debris or

The Ten-Tec model 580 was one of the first transceivers to offer coverage of the new bands authorized by WARC-79. (Courtesy of Ten-Tec Inc.)

started a fire. In many power supplies series resistors are used to limit the maximum current the power supply can deliver. Time delays are provided to prevent the high voltage from being applied to a partially heated vacuum tube. High-temperature cutouts are often provided. Never, under any circumstances, disable or delete these safety devices.

2.8 TRANSMITTER SPECIFICATIONS

Transmitters may be broadly classified by power capability and emission. A typical classification would be a medium-power CW transmitter or a high-power SSB transmitter. Emissions allowed on the high-frequency bands are CW, AM, FM, slow-scan television, and RTTY. Full carrier double-sideband AM is seldom, if ever, used. Some SSB transmitters, especially older designs, have AM capability.

Practically every SSB transmitter is capable of operating on CW. Also, most SSB transmitters are capable of operating on RTTY and slow-scan television using the appropriate external modulator.

The power capability may be specified as an input or an output, but must be specified for each mode of operation. Because of the different duty cycles of each mode of operation, the power capability will vary among emission types.

Single-sideband power is specified as a *peak envelope power* (PEP). This refers to the maximum instantaneous input or output occurring during a typical speech transmission. The short-term average power is approximately one-half of the PEP. Because of the pauses in normal speech, the long-term average power is less than one-half of the peak envelope power.

In CW emissions the PEP and the short-term average power are the same. However, the long-term average power is somewhat less because of the duty cycle of the Morse code.

Some SSB transmitters are not capable of operating CW at the maximum PEP value and must be detuned. Most good-quality SSB transmitters are capable of being operated at the PEP value for normal CW operation.

Radio teletype and slow-scan television put a lot of strain on most SSB transmitters. This is because the peak envelope, short-term average, and the long-term average are all the same. These modes of operation require the transmitter to supply carrier power for long periods of time, often for more than 10 minutes, without any off-time for cooling. In some transmitters, the maximum permissible power for RTTY or slow-scan television is much lower than the maximum PEP for SSB. Some manufacturers supply a fan for forced-air cooling and allow nearly the same power for RTTY and slow-scan television as is allowed for single-sideband PEP.

Historically, transmitters are rated for power input since the FCC regulations for maximum legal powers are specified as input powers.

Since the efficiency of power amplifiers varies considerably among amplifier types, different transmitters having the same input power rating may have significantly different output powers.

Efficiency, although not usually specified, may be calculated from the input and output powers by the formula

$$\text{efficiency} = \frac{P_{\text{out}}}{P_{\text{in}}} \times 100\%$$

Knowing the efficiency is helpful in selecting a transmitter for RTTY, slow-scan television, and CW operation. Typical efficiencies for class B linear amplifiers are on the order of 50 percent and often as low as 40 percent. Linear amplifiers must be used for SSB but are not required and are not desirable for RTTY, slow-scan television, and CW. Class C amplifiers, which are suitable, are between 60 and 75 percent efficient. Most SSB transmitters do not change the bias in the power amplifiers for CW operation. This low efficiency aggravates the cooling problems associated with RTTY and slow-scan television.

Frequency stability is a specification important to all emission types. Stability is usually specified as the drift in kilohertz per unit of time, after a specified warm-up period.

Dial accuracy is the ability to set the transmitter on a frequency by use of the dial or digital readout alone. Crystal calibrators are difficult to apply to a transmitter. Therefore, a transmitter dial accuracy specification may be different from a receiver specification, which usually assumes that the dial is first calibrated to the nearest crystal checkpoint.

Spurious output compliance is mandated by FCC rules and regulations. There are three types of spurious outputs: harmonics, images and other leakages, and intermodulation. Harmonics are usually generated in the power amplifiers, and can be eliminated by placing low-pass filters between the final amplifier and the transmitter output. Images and other leakages are generated within the transmitter and are eliminated by using balanced mixers, adequate filtering, and shielding. Intermodulation is generated in the linear RF power amplifier. Intermodulation in a single-sideband transmitter appears as splatter on the adjacent channel. The only way to eliminate this type of spurious is to maintain good linearity in the linear amplifier stages in the transmitter.

Other transmitter specifications are usually associated with the features and accessories, such as VOX times, ALC performance, and so on.

3 VHF AND UHF TRANSCEIVER FUNDAMENTALS

3.1 INTRODUCTION

Amateur activity in the VHF spectrum has made many major changes over the history of amateur radio. During the early days of amateur radio, absolutely no activity took place on those frequencies known today as the VHF spectrum. The vacuum tubes and components in use during that time were incapable of amplifying or oscillating at such a high frequency. It was not very long before the more clever amateurs were modifying vacuum tubes by removing their bases to reduce the series inductance of leads and altering other components for use on the VHF bands. Most of the early receivers were superregeneratives, while the transmitters were amplitude-modulated oscillators that probably had more frequency modulation than amplitude modulation. The superregenerative receivers were so wideband that the undesired frequency modulation made no difference to the intelligibility of the received signal. Amplitude modulation was the only type of modulation heard on the VHF bands except for an occasional FM or CW station. This continued for many years, up to the mid-1960s, when a lot of very inexpensive surplus commercial equipment was being made available by the FCC as the result of a major restructuring of the commercial land mobile bands. Amateurs were putting these rigs on the high end of the 2-m band for fixed-frequency point-to-point communications. These transceivers were installed without modifications other than the change to a noncommercial frequency. The original ±15-kHz wide-band modulation was retained, which was legal on the amateur band and was being phased out on the commercial frequencies. National calling frequencies were created and any amateur, regardless of how far from home, could find someone to talk to on these frequencies.

It was about this time that a new type of radio communication appeared on the 2-m band and won immediate acceptance. This was the introduction of the VHF FM repeater. From the late 1960s until the early 1970s, the 2-m band became the most active of all the amateur bands, and most of that activity was repeater operation, while amplitude modulation all but died. The wide-band modulation originally used by the early transplanted commercial rigs died rapidly and was replaced by a 5-kHz-deviation narrow-band FM. Also, activity increased on the 220- and 450-MHz bands, with repeaters appearing on those frequencies as well as SSB and CW.

The equipment used by amateurs also changed very rapidly during these times. Popular on the VHF bands for the 20 years preceding the explosive growth of FM activity were AM transceivers, usually with a dual (ac/low-voltage dc), power supplies that could be used for both mobile and fixed station use. Power inputs were in the 5- to 10-W range and were, of course, completely vacuum-tube. The more serious VHF operator would use a receiving converter in conjunction with a good-quality communications receiver for reception and a VHF transmitter in the 25- to 100-Watt category for transmission. The first amateurs to use FM used the converted taxicab- or police-band radio simply retuned for the 2-m band without modification and installed in an automobile as any commercial installation. It took quite a while before the manufacturers of amateur radio equipment recognized this growing market and produced a made-for-amateur radio 2-m FM transceiver. These units were either all solid-state or a mixture of tubes and transistors, crystal-controlled, and made for mobile use only. Operation as a base station required a low-voltage dc power supply or battery. As the number of repeaters grew, the band became practically saturated with repeaters on almost every frequency pair in the larger metropolitan areas. The cost of a large number of crystals made it economically feasible to provide a frequency synthesizer that could provide as many as 800 channels. Methods of preventing harmful interference between co-channel repeaters involved the transmission of tone bursts, and continuous tones called PL, after a commercial radio term.

The handy-talky, a light self-contained portable transceiver, was often used in commercial installations and began appearing on the surplus market at attractive prices when the commercial bands were restructured. Some of these units were also converted to 2-m FM service. These early units were hybrid tube and transistor affairs that had a very short battery life in normal amateur use. Many of the manufacturers that introduced 2-m FM transceivers for mobile use anticipated the market for completely portable units as well. These early amateur portable units were two- or three-channel crystal-controlled and had a power output of about 1 W. These units were replaced with higher-powered units with frequency synthesizers and other valuable features, such as tone squelch and signaling.

The discussion of VHF equipment will deal mostly with crystal-controlled and synthesized 2-m transceivers. Most of the activity on the 6-m band is on SSB and techniques used on the high-frequency (HF) bands are valid at the 6-m band. Equipment used on the higher-frequency bands, such as 220 and 450 MHz, are very similar to 2-m equipment except that the frequencies involved are proportionately higher. Special weak-signal work, such as moon bounce and meteor scatter, involves sophisticated equipment, discussion of which is beyond the scope of this book.

Another unit unique to the VHF spectrum is the transceiving converter. This unit converts the received signal from the antenna to a lower-frequency amateur band and converts the transmitter signal from that same low-frequency band up to the VHF band. Since the same local oscillator is used to convert both the transmit and receiver signals, the converted signal is translated by exactly the same amount. This allows for a high-frequency transceiver to be used on the VHF bands without any loss of flexibility. The frequency converters are linear, so that whatever signal is fed into the converter emerges unchanged at the VHF band. That is, if SSB is fed into the converter, the output signal has exactly the same modulation with a minimum of added distortion. This system is analogous to adding another stage of conversion to a dual- or single-conversion superheterodyne receiver or to a heterodyning type of transmitter.

3.2 VHF RECEIVERS

The requirements for a receiver in the VHF spectrum are the same as those in the high-frequency spectrum. All VHF receivers supply gain, selectivity, and demodulation, even though the techniques used are somewhat different from those previously discussed. In the VHF spectrum, the sensitivity and noise figure of the receiver are very important. Unlike the high frequencies, there is very little atmospheric noise to reduce communications ability, which implies that the ability to communicate is mostly a matter of transmitter power and receiver sensitivity.

In the high-frequency receiver, often an RF amplifier or active mixer were not used, in favor of connecting the antenna directly to a passive balanced mixer. This greatly improves the dynamic range of the

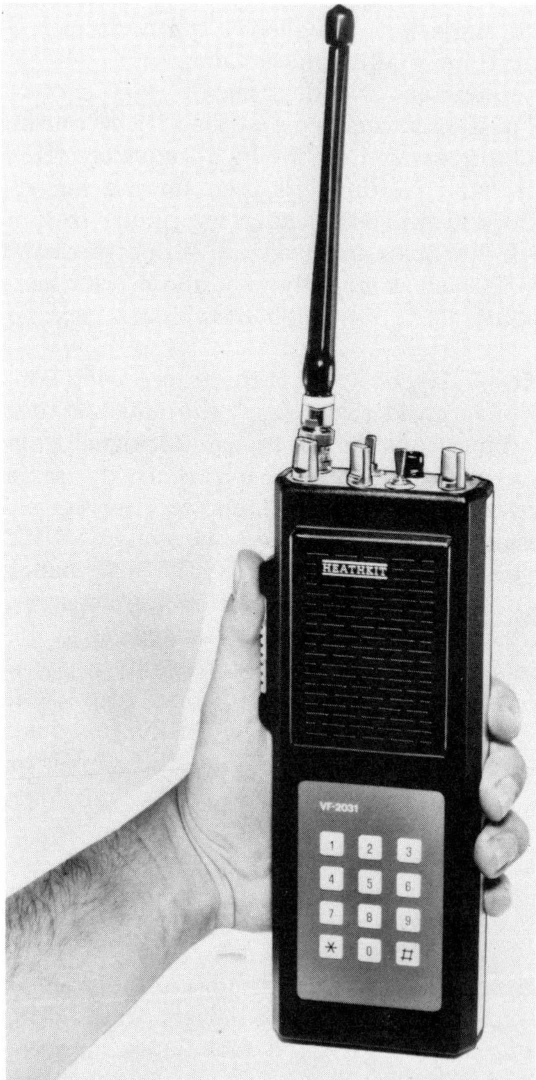

The Heath VF-2031 is a hand-held 2-m transceiver available in kit form. (Courtesy of Heath Company.)

receiver but seriously affects the noise figure. In the VHF spectrum, especially on 144 MHz and higher, a noise figure of 2 dB or better is common and desirable. This noise figure cannot be obtained without a good-quality RF amplifier. Aside from this and the generation of a stable local oscillator, as with a synthesizer, the VHF receiver is very much like its high-frequency counterpart.

3.2.1 FM Receivers

Limiters. VHF FM receivers are the most common and are simple examples of VHF receiver design. They are usually single-conversion superheterodynes with a good-quality crystal filter for selectivity followed by several stages of limiting and amplification plus the usual demodulation. The entire concept of limiting is unique to FM and contributes to some of the more useful characteristics of FM modulation. A limiter amplifier is one in which, beyond a certain input level, there is no further increase in output level. This implies that the amplifier is saturated. In an FM IF amplifier, several limiting amplifiers are cascaded togeher so that the last limiter stage reaches saturation when the input is on the order of several microvolts. The last stage limits first, followed by the next-to-last stage, the next, and so on, until the first limiter stage has saturated. The change in imput signal level from the level the last limiter stage saturates until the first limiter saturates can be 60 dB or greater. Smooth-operating limiting amplifiers with a large dynamic range are required for proper FM receiver operation. The ability of the limiters to handle a large dynamic range obviates the need for an AGC system, except for an occasional use of AGC on the RF amplifier to prevent overload.

Noise on weak signals appears as amplitude variations on the carrier. The action of the limiting amplifier is to remove or greatly attenuate any amplitude variations and thus effectively remove most of the noise. The amount of noise reduction depends on the amount of limiting taking place and is mainly dependent on signal strength. This phenomenon is called "quieting" and is an indication of signal strength.

The block diagram of a simple FM receiver shown in Figure 3–1 includes, as well as the usual mixer, selectivity, and IF amplifiers, a block called *squelch*. In a properly operating FM communications receiver, a loud hissing noise is present whenever no signal is being received. To remove this annoying noise, an electronic system removes the audio input from the audio amplifier whenever no signal is being received and allows the audio to be restored whenever a signal is present.

Squelch Systems. Figure 3–2 shows the schematic of a squelch system. In this sytem, high-frequency noise voltage is rectified by

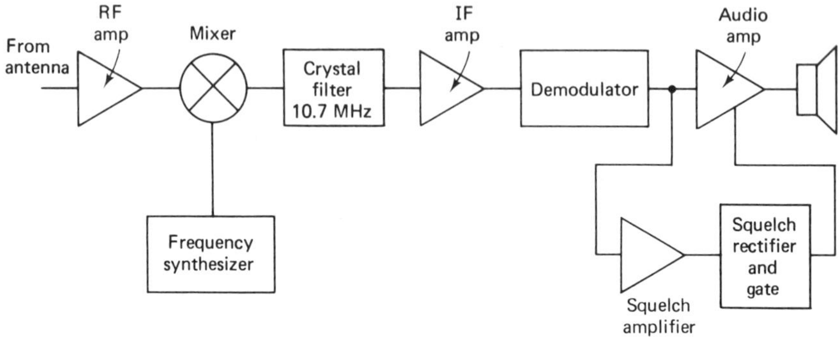

FIGURE 3-1. Single-conversion FM receiver using a frequency synthesizer.

diodes D1 and D2 and applied to comparator A1. Usually, the noise voltage is high-pass-filtered to prevent interference from the demodulated voice frequency below 3 kHz. The output of the comparator changes the state of an electronic switch that disconnects the input of the audio amplifier. Even very weak signals will reduce the amplitude of the background hiss and cause the squelch to activate or "open." The ability of an FM receiver to effectively quiet with a weak signal is a function of the limiting amplifiers and is also a measure of its sensitivity.

As with high-frequency receivers, the image differs from the desired frequency by twice the IF. Therefore, the higher the IF, the more the

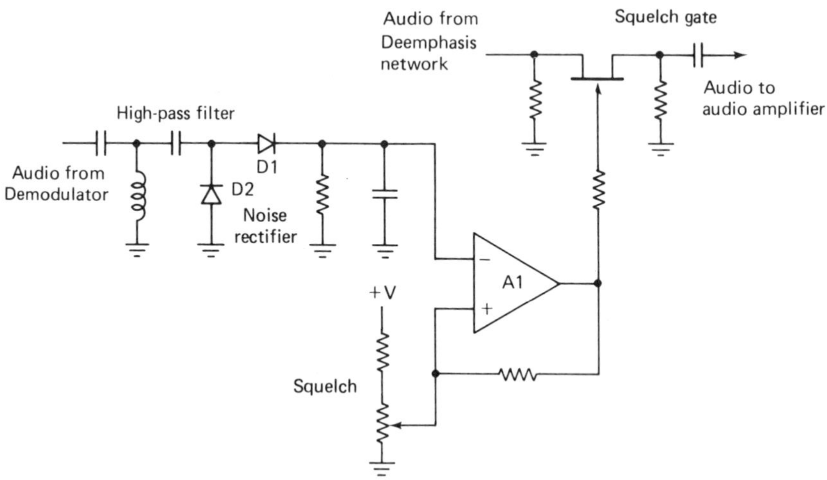

FIGURE 3-2. Simple squelch system using a FET for an audio gate.

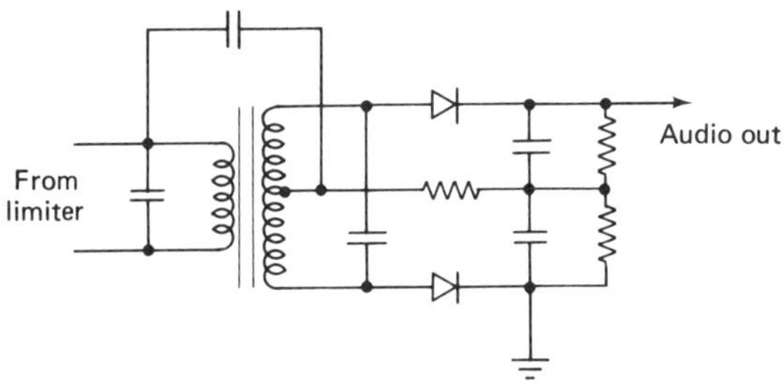

FIGURE 3-3. The frequency discriminator.

reduction of the image by the front-end filter. A popular IF for VHF receivers is 10.7 MHz, with 21.4 and 45 MHz becoming popular, especially for 220 and 450 MHz. Monolithic crystal filters sufficiently narrow for FM reception are available for all the popular IFs and are available at attractive prices. Filters for the more narrow bandwidth modes, such as CW and SSB, are not generally available at these higher intermediate frequencies.

Detectors. Detector systems for FM reception make use of the phase-shift characteristics of a tuned circuit at or near resonance. The most popular detectors are the discriminator, shown in Figure 3-3; the ratio detector, shown in Figure 3-4; and the quadrature detector, shown in Figure 3-5. The discriminator and the ratio detector require special IF transformers for their construction, whereas the quadrature detector uses a single tuned circuit. Although the quadrature detector has been used for many systems, especially television receivers, it is being used more often in high-quality communications systems because of the availability of integrated circuits. Excellent-performing narrow-

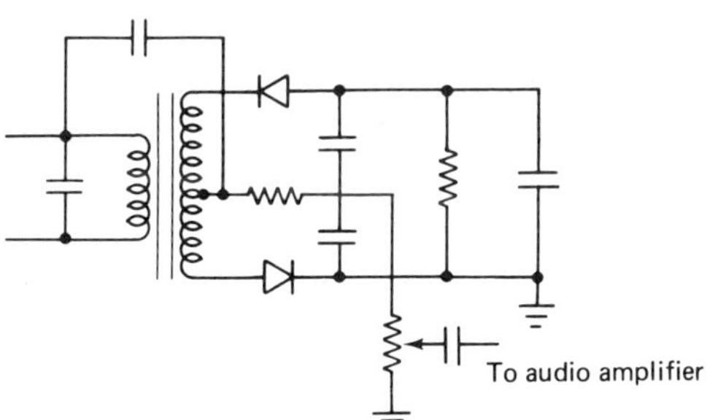

FIGURE 3-4. The ratio detector is very similar to the frequency discriminator but offers improved performance.

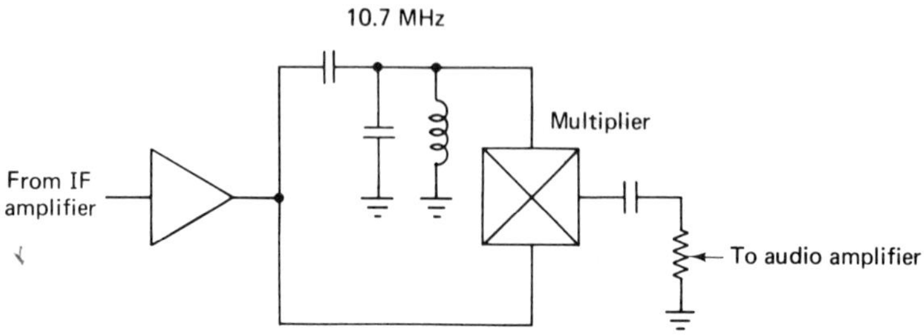

FIGURE 3-5. The quadrature detector is a popular circuit for implementation with integrated circuits.

band detectors are constructed with a quadrature detector IC and a monolithic crystal filter.

Referring to the block diagram of the quadrature detector in Figure 3-5, the IF signal, after amplification and limiting, is divided into two separate paths. One path goes directly into an input of a multiplier, while the other drives a tuned circuit that feeds the other input of the multiplier. This sort of multiplier is not to be confused with the frequency multiplier used to generate harmonics of a signal. This multiplier performs the mathematical function of multiplying two signals together and is similar to a mixer. Since both of the multiplier input signals are of exactly the same frequency, the resultant output is a dc signal that is proportional to the sine of the phase angle between the two signals. Since one path was through a tuned circuit while the other path was not, the output voltage is a function of the phase angle between the two signals, which varies with frequency depending on the Q of the tuned circuit. High-Q circuits will provide a large amount of recovered audio, while lower-Q circuits are used for large-deviation signals such as commercial broadcast. For high audio recovery with narrow deviations, extremely high Q is required. Coupled with the requirement for stability over temperature and aging, this places stringent requirements on the tuned circuit used for high-quality communications receivers. Special monolithic crystal filters have been adapted to serve as a tuned circuit for the quadrature detector used in 2-m FM equipment that offers high recovered audio with excellent temperature tracking and small size.

Another type of detector that has found use in dual-conversion receivers where the second IF is a low frequency, such as 455 kHz, is the *phase-locked loop*, as shown in block diagram form in Figure 3-6. The phase-locked loop is designed with a relatively high loop bandwidth, compared to the modulation bandwidth, so that it can follow deviations in the input frequency due to voice-frequency modulation. As the loop tracks the input frequency, the VCO control voltage will

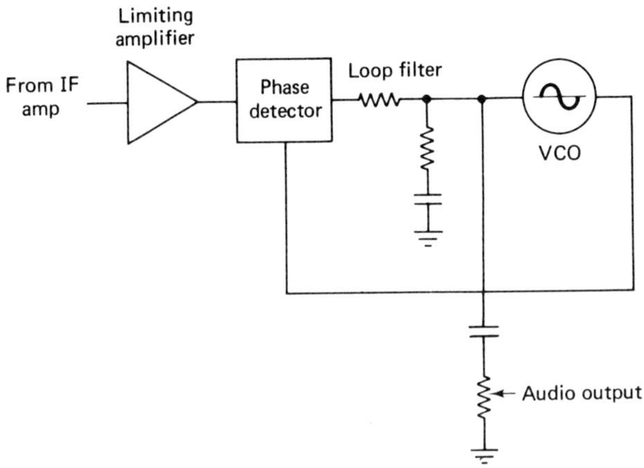

FIGURE 3-6. The phase-locked-loop frequency discriminator is best suited for low frequencies.

vary at the modulation rate, which can be amplified for use as a demodulated output. The major advantage of the phase-locked-loop demodulator is that no coils are required. This is especially important at low frequencies, where small high-Q inductors are difficult, if not impossible, to manufacture. Good sensitivity without limiter amplifiers is possible with many integrated-circuit phase-locked-loop systems.

3.2.2 Receivers for Other Services

Receivers in the VHF spectrum for single-sideband and other services, such as CW, differ very little from their high-frequency counterparts. Figure 3-7 shows a single-sideband receiver for the 50-MHz band. It can be seen that the circuits and functions of all the blocks are no different from those found on many classical dual-conversion designs for the lower frequencies. This is usually the case for SSB equipment on the 6- and 2-m bands, but on 220 MHz and above, more exotic equipment is the rule. Very little equipment is available for sideband use from manufacturers at these frequencies, and most amateur work is for distance records and satellite communications. A receiver for satellite or other weak-signal work is usually an exotic front end followed by a normal high-frequency communications receiver. Most high-performance front ends of the exotic kind are dependent on high-performance, expensive semiconductor devices such as gallium-arsenide FETs and specially constructed bipolar transistors. The techniques unique to 450 MHz and even 220 MHz and higher are beyond the scope of this book and will not be discussed.

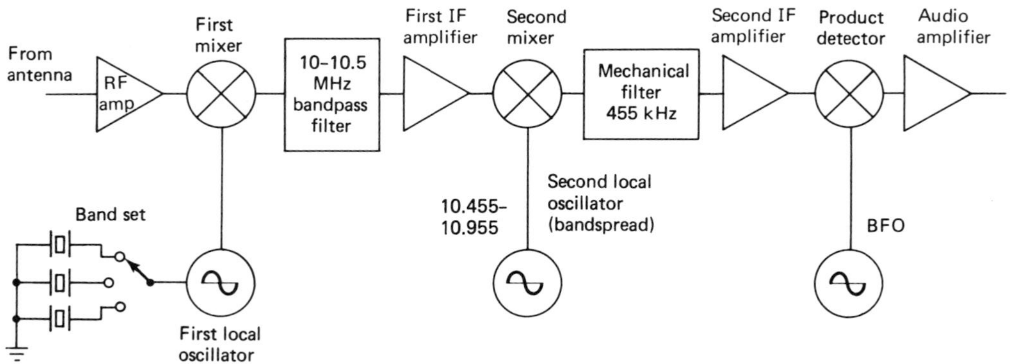

FIGURE 3-7. Dual conversion with a crystal-controlled first local oscillator is typical of VHF single-sideband receivers.

3.3 VHF FM TRANSMITTERS

3.3.1 Modulators

The simplest VHF transmitter uses a crystal oscillator and multiplier stages to reach the desired output frequency. This technique has been in use from the days of the very first crystal-controlled VHF AM transmitters, up to the present time in the form of inexpensive solid-state FM equipment. As shown in the block diagram of Figure 3-8, a crystal oscillator, operating at a fraction of the output frequency, is multiplied by frequency multipliers to the output frequency. Modulation may be applied to the crystal oscillator by placing a varactor diode in series with the crystal and modulating the varactor diode with the audio-frequency modulation. This method works best with a fundamental mode crystal which limits the maximum frequency of the crystal to about 20 MHz. Since 7 is not an easy multiplication, the closest practical multiplication would be 8, which would use crystals in the 18-MHz range. For a single-channel transmitter, this method would provide adequate modulation for any amateur band above about 144 MHz. The major disadvantage is that the amount of modulation may depend on the crystal used. Some crystals will give considerably more modulation than others, which becomes a serious problem when a number of crystals are used.

A frequency-modulated crystal oscillator is shown in Figure 3-9. Since the actual operating frequency of a crystal oscillator depends, slightly, on the reactance that is placed across the crystal, slight variations in frequency may be achieved by placing a varactor diode in series with the crystal and modulating the varactor at the audio-frequency rate. The amount of capacitance that a crystal is to be operated into is specified when the crystal is manufactured and is usually in the 10- to 47-pF region. Varying the capacitance around the nominal value will cause the operating frequency of the crystal oscillator to vary around the nominal frequency. Usually, no more than 0.1 percent of frequency modulation may be obtained in this way before severe nonlinearities set in. For narrow-band FM, 50 MHz is the lowest-frequency amateur band that may use this form of modulation.

Using a phase modulator rather than modulating a crystal oscillator will provide lower distortion and a constant modulation level regardless of the crystal used. However, phase modulation must be converted to frequency modulation by adjusting the frequency response of the modulating signal so that the lower frequencies are enhanced. Figure 3-10 shows a phase modulator following a crystal oscillator suitable for narrow-band frequency modulation in the VHF spectrum.

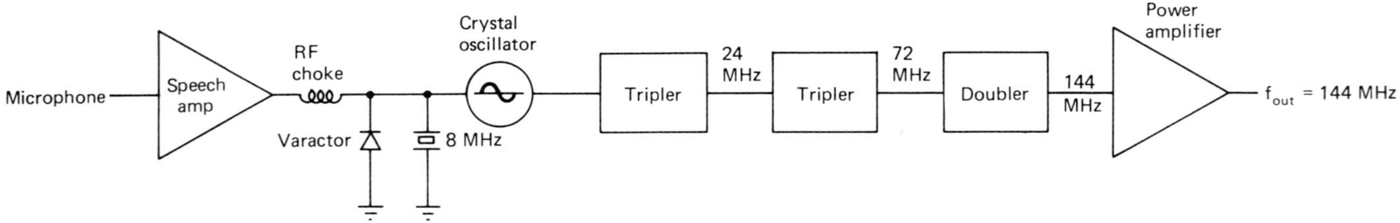

FIGURE 3-8. Crystal-controlled FM transmitter with modulation applied directly to the crystal oscillator.

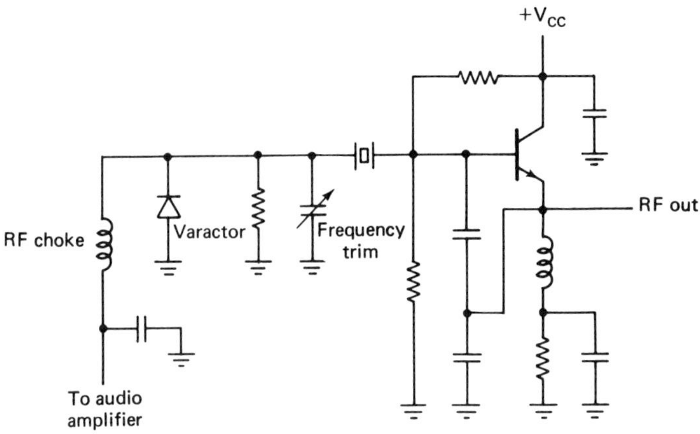

FIGURE 3-9. Direct FM crystal oscillator using a varactor diode.

In this circuit an FET amplifier stage acts as a variable-gain RF amplifier while some of the input signal is fed around the amplifier through C2. As the gain of the phase modulator is varied by modulating the bias of the FET amplifier by the application of an audio-frequency modulating waveform, the output will receive signals from either the amplifier, which is phase-reversed, or directly through C2. This causes the phase difference between the input and the output of this circuit to

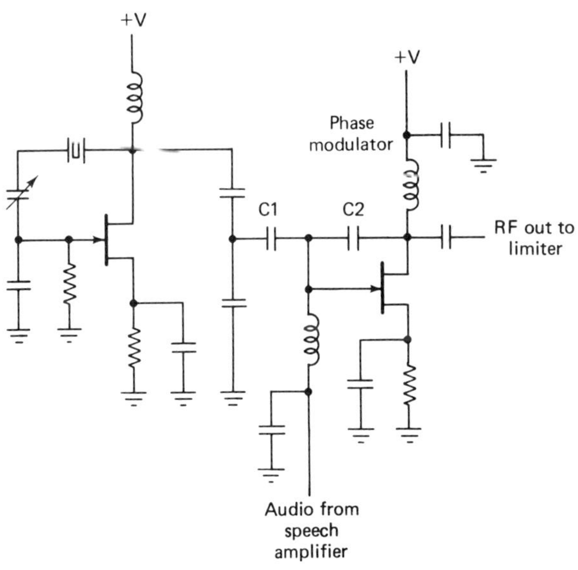

FIGURE 3-10. Phase modulator following a Pierce oscillator.

vary proportionally with the modulation amplitude. Usually, a phase-modulated FM transmitter will require more frequency multiplication than a transmitter that is direct-FM-modulated with a varactor diode. For a 5-kHz frequency swing at the 2-m band, the minimum multiplication required is about 12 times.

Every phase modulator will produce a small amount of amplitude modulation, which should be removed before frequency multiplication. A limiter stage, similar to those used in FM receivers, is inserted after the phase modulator and before the first frequency multiplier, to remove any trace of amplitude modulation.

3.3.2 Frequency Multipliers

A frequency multiplier is simply an overdriven amplifier whose output is tuned to a harmonic of the input frequency as shown in Figure 3-11. Schematically, a frequency multiplier appears to be no different from any other tuned amplifier. The things that make it a multiplier are the drive level, which is very high, and the frequency of the output tuned circuit, which is not the same as the input frequency. It is a good idea to have an active device with a high-frequency capability for efficient frequency conversion. In modern transmitters, spectral purity is a must, which requires the output spectrum of a frequency multiplier to be clean. It is often necessary to use a high-loaded Q as well as multiple tuning to remove all unwanted harmonics. Vacuum-tube transmitters did not suffer from this problem and, to a certain extent, FET multipliers do not either. Single tuning was used primarily in older vacuum-tube multiplier-type transmitters.

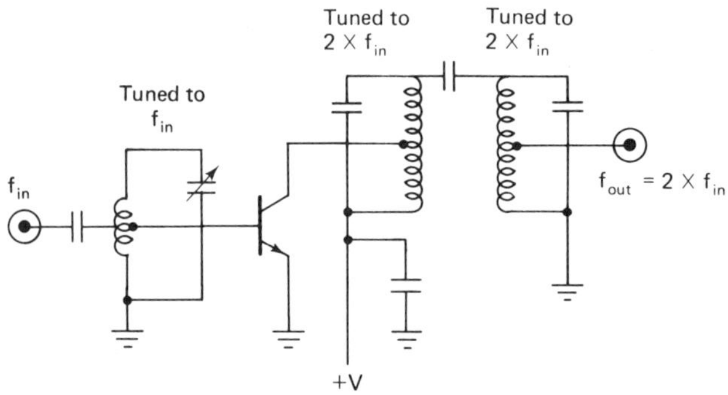

FIGURE 3-11. Frequency doubler with a double tuned output for spectral purity.

3.3.3 Crystal-Sharing Circuits

Normal receiver or transmitter design would require a separate crystal for both the transmitter and receiver. With transceivers operating on a large number of channels, especially portable transceivers, in which space is at a premium, it is desirable to reduce the number of crystals required. There are several methods of reducing the number of crystals per channel to one, and the majority of these schemes use a crystal to supply the receiver local oscillator and heterodyne that frequency with a crystal at the IF for simplex and other frequencies for standard repeater offsets.

Figure 3-12 shows a block diagram of a typical crystal-reducing scheme. This system is used in a hand-held transceiver and allows eight channels to be installed within a relatively small case. The crystal frequency is one-ninth of the receiver local oscillator frequency and is multiplied and supplies the receiver local oscillator requirements. The local oscillator signal is heterodyned with 10.7 MHz for the simplex transmit frequency and with 11.3 MHz for a transmit frequency 600 kHz above the receiver frequency and 10.1 MHz for a transmit frequency 600 kHz below the receive frequency. The output of the mixer must be very carefully filtered to remove completely any trace of the receiver local oscillator signal that is only 10.7 MHz removed.

A VFO could be used with a frequency-multiplier type of transmitter instead of crystals, which was often done during the AM days of VHF operation. For FM operation with repeaters, the VFO would lack the stability and dial accuracy to be of any use. To provide frequency control, without the restrictions of crystals and with the stability required, the modern FM transmitter uses a frequency synthesizer, which lends itself well to the channelized operation of 2-m FM.

3.3.4 Synthesizers

The synthesized FM transceiver is unique to the VHF spectrum and is unlike any other equipment discussed previously. The synthesizer dictates the performance of a transceiver in several areas and often dominates the circuitry of a VHF transceiver. There are as many different types of synthesizers as there are manufacturers, and for the purposes of discussion, they will be divided into two basic categories: direct, on-frequency synthesis and feedback mixing. Direct, on-frequency synthesis, although the simplest-appearing system from the block diagram shown in Figure 3-13, is the least used. The major dis-

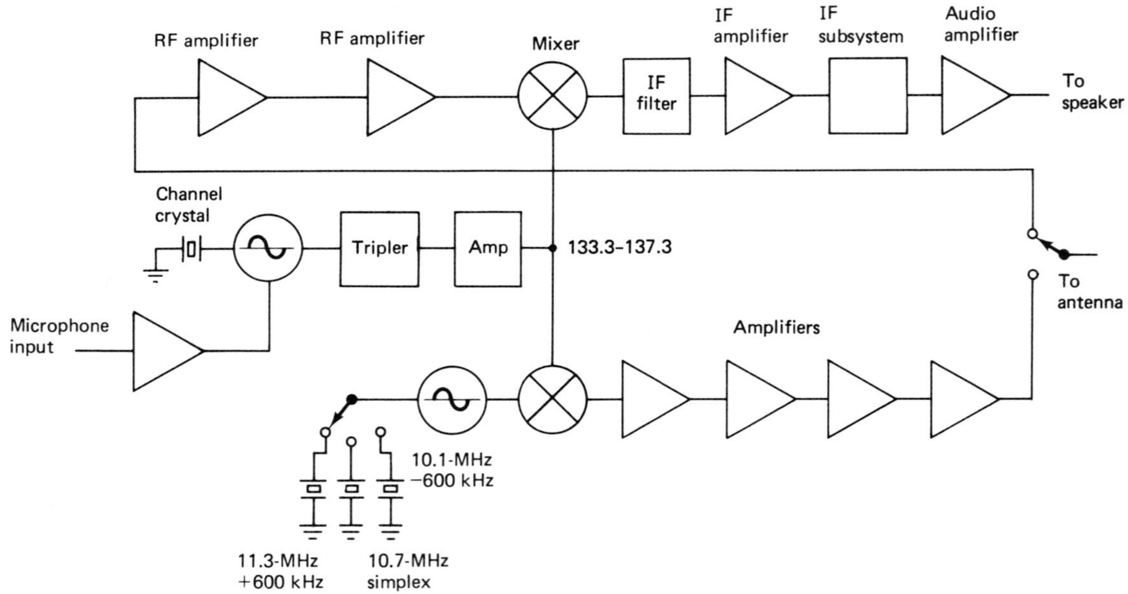

FIGURE 3-12. Method of reducing the required crystals to one per channel in a 2-m handy-talky (Heath Model VF2031).

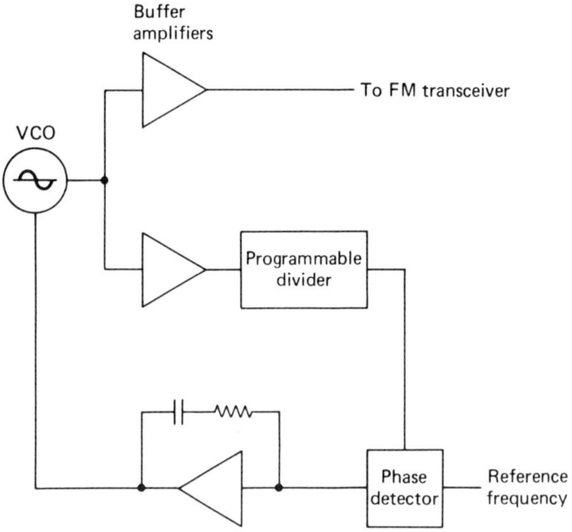

FIGURE 3-13. Basic frequency synthesizer using a phase-locked loop.

advantage of this system is the use of a very high speed digital prescaler required for the programmable divider, which is not only expensive but usually consumes considerable power from the supply. A synthesizer used in a transceiver must supply not only the proper transmitter frequency, but a local oscillator for the receiver as well. The frequency of the local oscillator must be different from the actual received frequency by the receiver's intermediate frequency. In addition, the selector switches must indicate the actual receiver frequency and not the local oscillator frequency. In the case of repeater operation, the offset will include the repeater offset frequency as well. Summing it all up, the synthesizer frequency is not always the same as that read on the transceiver dial. This requires some sort of manipulation of the programmable divider, which, in the case of the direct synthesis system, must be done digitally within the programmable divider logic or with a microprocessor.

The on-frequency direct synthesizer shown in Figure 3-13 contains only the basic blocks and is a good example for understanding the phase-locked-loop synthesizer.

The fundamental idea of the phase-locked-loop synthesizer is to use a voltage-tuned oscillator that is capable of covering a relatively wide frequency range and stabilizing its frequency by comparing a fraction of that frequency to a very stable signal called the reference. For example, assume that 147.015 MHz is desired and that a 5-kHz reference is being used. Thus, the programmable divider is set to 147,015 divided by

5, resulting in 29,403. That is, the output frequency, which can only be a multiple of the reference, will be 29,403 times the reference frequency. If the voltage-tuned oscillator is operating at precisely 147.015 MHz, the output from the phase detector will be such that no change will occur in the VCO frequency. If the frequency should drift higher or lower, or be off the desired frequency for any reason at all, the output from the phase detector will be such as to correct for the frequency error. This process operates very quickly, so that the circuit can correct for slow changes induced by temperature changes and rapid variations, such as microphonics, which are variations in frequency due to the mechanical vibration of parts in or near the VCO, and phase noise, which are rapid random variations due to noise in the oscillator itself. The phase detector operates at a frequency of 5 kHz and can sample the phase difference only at this rate. This implies that corrections can be made only at the reference frequency. If the corrections were fed back without any filtering, very strong modulation would be applied at the 5-kHz rate. Therefore, it is necessary to filter the correction voltage to remove most of the 5-kHz signal and retain only the average voltage to be used as a correction voltage. For the cleanest signal, the reference frequency should be as high as practicable.

In Figure 3–13, the voltage-controlled oscillator is usually a varactor-tuned oscillator using a field-effect or bipolar transistor that tunes the desired band of frequencies. The reference signal source is usually a crystal oscillator and several stages of digital frequency division. The phase detector and the loop amplifier are common integrated circuits created especially for phase-locked-loop circuits. It is the programmable divider that is unique to this type of VHF synthesizer. For a programmable divider to operate above about 25 MHz, a technique often used is dual modulus prescaling. The device that makes this system possible is an emitter-coupled-logic high-speed dual modulus prescaler integrated

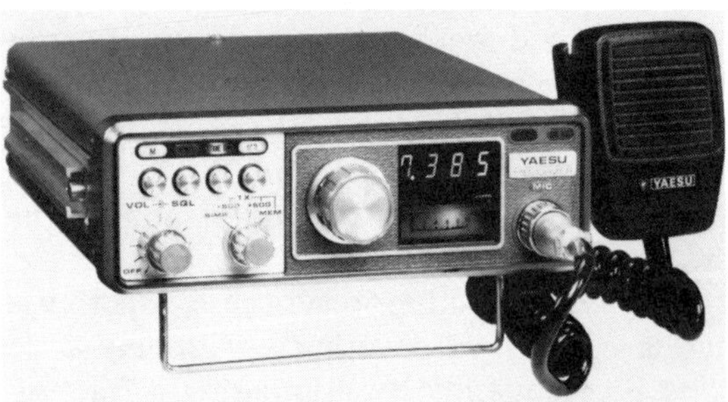

The Yaesu FT-227 is a synthesized 2-m FM transceiver with frequency memory capability. (Courtesy of Yaesu Electronics Corp.)

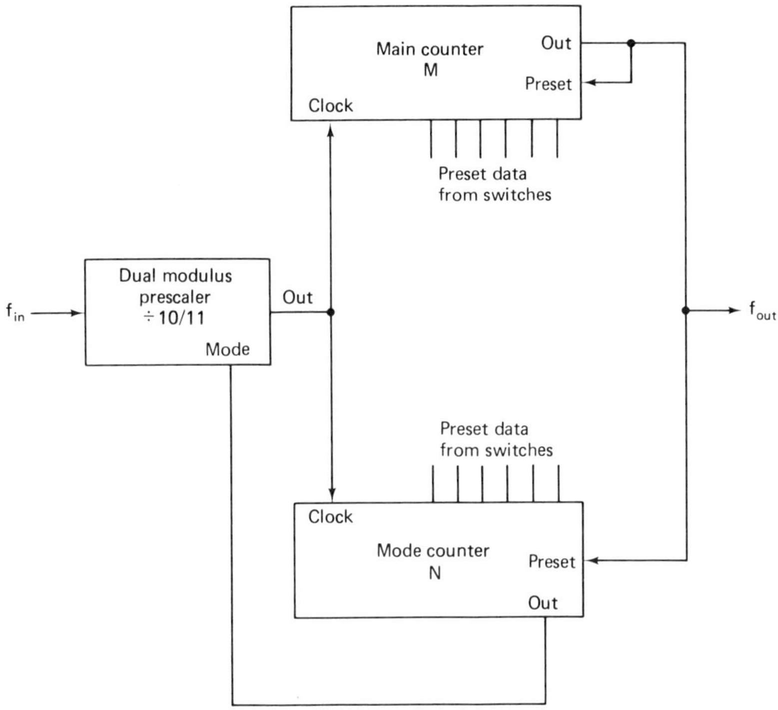

FIGURE 3-14. Dual-modulus prescaler used in a direct synthesis divide-by-N circuit.

circuit, which is capable of dividing an input frequency by one of two moduli. For example, a very popular dual modulus prescaler is the 10/11 divider. By setting a logic level at the divider's mode input, the chip will divide the input frequency by 10, and by changing the state of the mode input, the divider will divide by 11. Figure 3-14 shows how this chip is used to create a VHF programmable divider. Two down counters are used and preset at the same time. The *main counter*, called the *M counter*, is at the top, and the second counter, called the *mode counter* or *N counter*, is at the bottom. The top counter causes all the counters to be preset when it reaches zero. The mode counter also counts down to zero, but unlike the main counter, does not continue to count until the main counter has reached zero. Upon reaching zero, the mode counter changes the mode from the higher divisor, usually 11, to the lower divisor, which is usually 10, and keeps the mode in this state until the main counter has reached zero.

Consider the number of input pulses required to complete one cycle of the programmable divider. The number of output pulses from the dual modulus prescaler required for the complete cycle is exactly equal to the number preset into the main counter, M. If the prescaler divided only by 10, the total number of pulses would be $10M$. But this is not the case. For part of the cycle, that is, the time before the mode counter reached zero, the dual modulus prescaler was dividing by 11, and for the remainder of the output pulses, was dividing by 10. Therefore, for M pulses, the division was 11 and for $M - N$ pulses, the division was 10. The total number of input pulses is equal to $11N + 10\ (M - N) = N + 10M$. It can be seen from this result that the number M has 10 times the effect of N on the programmable divider. Another restriction is that M must be greater than N. Figure 3–15 shows a programmable divider using decade counters for the main counter and a binary counter for the mode counter. One-megahertz and 10-MHz data are supplied to the main counter while tens of kilohertz and the 5-kHz offset switch are applied to the mode counter. These data could be supplied from coded binary switches if the divider was to be used in a transmitter. However, since most synthesizers are used in transceivers, where the receiver frequency is usually displayed, the displayed frequency and the actual operating frequency of the synthesizer differs by the IF of the receiver. This subtraction is not easily done by simple logic but should be accomplished by a microcomputer, which could handle other tasks, such as switching between transmit and receive, memorizing channels, scanning, and other chores. Other techniques, such as presetting the counters on both the M and N counters before reaching zero, can generate the proper offset frequency. The amount of logic required and the lack of flexibility makes it desirable to install a small limited-function microcomputer.

Another problem that is unique to the direct-synthesis method is the problem of supplying modulation. The entire purpose of the phase-locked loop is to remove any variations in frequency caused by a multitude of reasons. Applying modulation is a purposeful application of frequency deviation that the phase-locked loop will try to remove. It is necessary to restrict the ability of the phase-locked loop to remove deviations at an audio-frequency rate, so that the VCO can be deviated by the modulation. Usually, this is accomplished by restricting the loop bandwidth to 300 Hz so that the ability of the loop to remove variations more rapid than a 300 Hz will be greatly attenuated. Once this is done, the VCO must be free of microphonics, whose mechanical vibrations are usually in the audio-frequency range.

Because of the expense of the microcomputer and the dual modulus divider integrated circuit, many other schemes have been devised to provide the receiver IF offset and to allow one synthesizer to be used for both transmit and receive. Most of these schemes involved mixing with one or more crystal oscillators. The IF offset required for the receiver local oscillator, plus any repeater offsets and even the 5-kHz

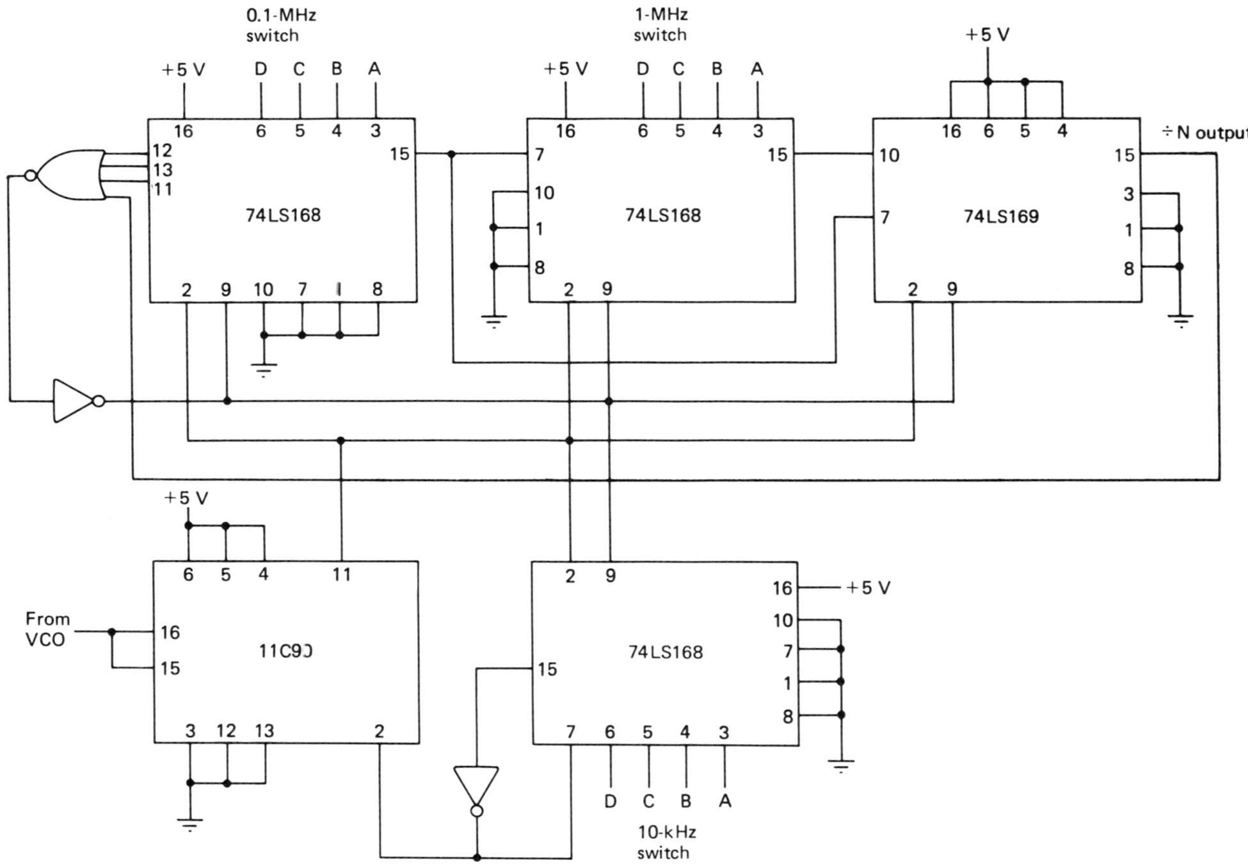

FIGURE 3-15. Dual-modulus divider using LS TTL and ECL logic. The maximum frequency of this circuit is above 150 MHz.

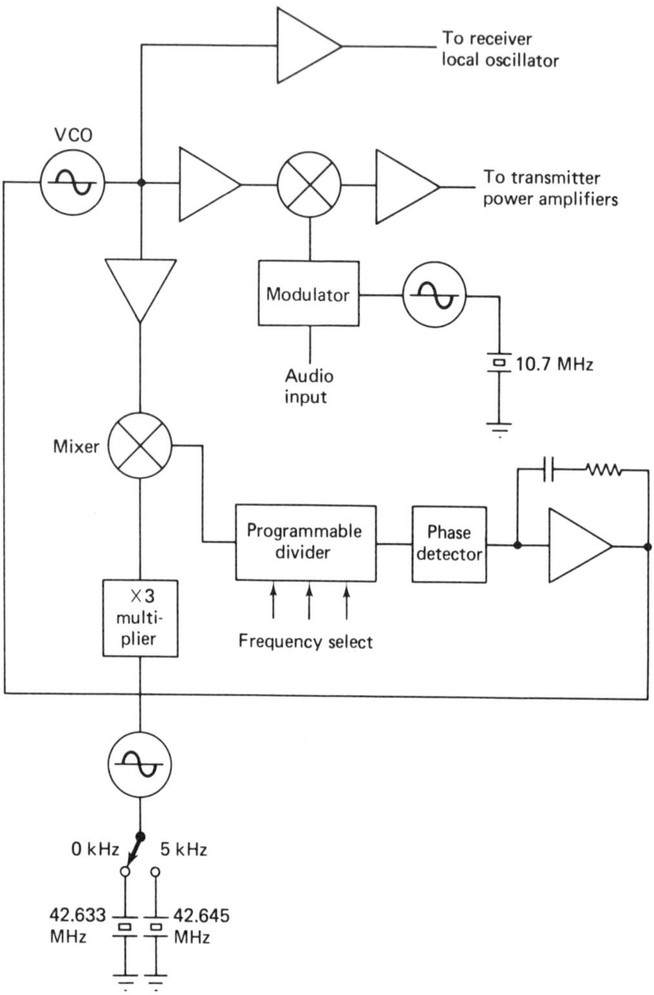

FIGURE 3-16. Using mixers to provide the transmitter offset and to lower the frequency to the programmable divider decreases the dependence on VHF logic.

shift, is obtained with crystals. The synthesizer shown in Figure 3-16 is a feedback-mixing type of circuit and is typical of a synthesizer used in a 2-m mobile transceiver. In this synthesizer, the VCO always operates at a frequency removed from the transmitting or receiving frequency by 10.7 MHz. To create the transmit frequency the VCO is heterodyned with a frequency-modulated 10.7-MHz signal. One advantage of this technique is that modulation is easy to apply and remains constant regardless of the VCO frequency. Another advantage is that the VCO frequency does not change very much between transmit and receive. For simplex operation, in which the transceiver transmits and receives on the same frequency, the VCO does not change frequency at all. If repeater operation is used, the VCO shifts only the amount of the repeater offset between transmit and receive, which is considerably less than 10.7 MHz, another popular IF. Large shifts in the VCO between transmit and receive require very careful phase-locked-loop design to ensure quick lockup between transmit and receive.

The VCO is mixed with a crystal oscillator before feeding the programmable divider. Two functions are provided by the mixing action. First, the VCO frequency is translated to a lower frequency that can be handled by a less expensive, slower type of logic than is required by the direct synthesis system. Second, the difference in frequency between the two crystals after multiplication is exactly 5 kHz, which supplies the required 5-kHz shift.

3.3.5 VHF Power Amplifiers

Increasing the power from a low level from the VHF synthesizer, sideband generator, or other stages in the VHF transmitter to the desired power output is usually accomplished by several stages of amplification. For FM systems, the amplifiers can be class C for maximum efficiency, while, as always, SSB service requires linear amplifiers all the way. The number of amplifier stages is determined by the power-output requirements and the power-input level. The higher the power level, the less gain is usually available from the transistor amplifier. At the low power level, that is, below 1 W, it is not difficult or unusual to achieve gains of 10 dB or more, while the higher-powered stages will have less gain, reducing to about 4 dB for very high-powered stages. Power levels up to about 50 W are available from a single transistor, while power levels greater than 50 W are usually obtained with multiple devices. For example, about 100 W would be available from two transistors operating in a push-pull mode, while four transistors operating as two push-pull amplifiers, feeding a power combiner, would be used for about 180 W. The efficiency of very high power amplifiers decreases due to losses in the power combiners. For extreme high powers, such as those used for broadcast transmitters and other long-distance communications services, vacuum-tube final amplifiers are currently being used.

VHF power amplifiers usually have very simple circuits, but in actual construction, must be very carefully made to avoid parasitics and other instabilities. All the problems of parasitic inductances and packaging capacities have a greater effect at VHF frequencies than at high frequencies. Special VHF packages are used by semiconductor manufacturers to facilitate construction of VHF power amplifiers. This usually involves multiple emitter leads, opposed base and collector leads, and very broad flat leads for reduced inductance. Usually, a manufacturer specifies a transistor for a certain supply voltage, operating at one power level in a well-defined frequency range. Transistors are tested at these parameters, and careful controls are placed on the production line to meet all the specifications. This means that surplus transistors that have no known specifications are very poor choices for home-constructed amplifiers. Most manufacturers of commercial communications equipment work very closely with the semiconductor-device supplier to create a circuit that will do the required job safely and effectively, and then both parties stick to their design.

Since the slightest lead length at VHF can cause a significant reactance, as compared to the circuit impedances, it is nearly impossible to use conventional components or wires for carrying circuit currents. VHF power amplifiers often use strip-line techniques for making broadband amplifiers, where energy is conveyed only by transmission lines printed on the circuit board.

A medium-power amplifier for 144 MHz is shown in Figure 3–17. The power level of this circuit is not so high that strip-line techniques must be used. As long as reasonable care is taken, conventional components may be used. In this amplifier, the 50-Ω impedance of the driver is matched to a lower impedance of the base with C1, C2, L1, and C3. In some circuits C3 is not used, but the matching network uses the internal capacitance of the transistor. A choke is connected from the base to ground to keep the base voltage at zero. A transformer and two capacitors are used for output circuit matching.

A higher-power amplifier using two devices is shown in Figure 3–18.

The valuable characteristics of MOSFETs have been known and used for many years for small-signal transistors in the ultra high frequency (UHF) spectrum. Recently, the power MOSFET has appeared as a workable device for generating power in the VHF and the HF spec-

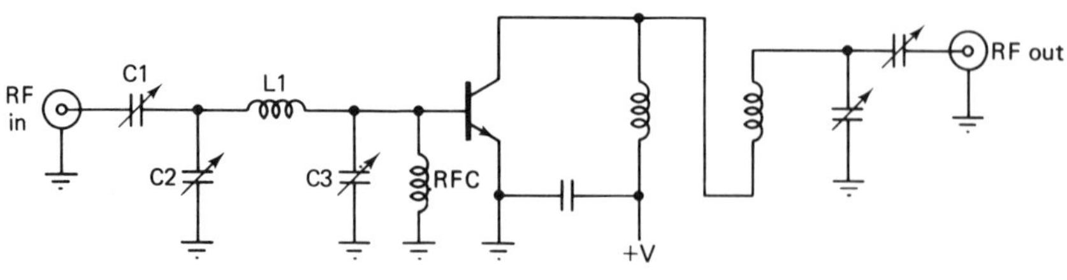

FIGURE 3–17. Wide-band transformer-coupled VHF amplifier.

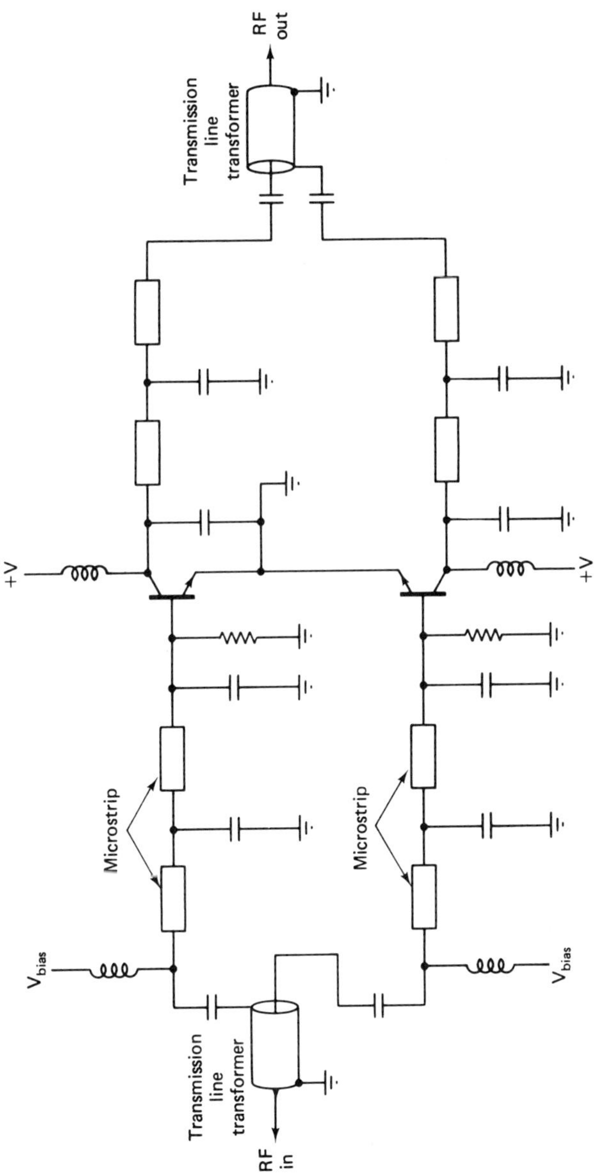

FIGURE 3-18. Push-pull linear amplifier for UHF. The inputs and outputs are matched with transmissior lines.

trum. Powers of over 100 W are currently being generated in both class C and linear amplifiers. Several types of MOSFETs are available in low-inductance packages, making them suitable for VHF operation. The power MOSFET amplifier can offer more gain at VHF frequencies than a typical bipolar power transistor and has superior intermodulation characteristics for linear amplifiers. It seems that the trend for higher-powered MOSFETS will be through higher voltages and lower currents rather than higher-current devices operating from low voltages such as 12 V. A typical power MOSFET amplifier is shown in Figure 3-19. Most of the construction techniques used for bipolar VHF power amplifier design are valid in this circuit. And the only real advantage to using such a circuit is for its excellent thermal stability and freedom from secondary breakdown and thermal runaway.

All the VHF power amplifiers described do not have the spectral purity to be used on the air without a low-pass filter to remove the harmonics generated in the power amplifier. FCC regulations require that the harmonic energy of a transmitter be 60 dB down from the level of the carrier. At least a two-section low-pass filter must be used to achieve this level of harmonic energy.

As in all solid-state amplifiers over the level of a few watts, the VHF power amplifier should be protected from the damage caused by excessive SWRs. This can be done with a power-cutback system that is activated by an SWR sensor. The cutback can be somewhat less severe in class C amplifiers used for FM service than the cutback required for linear amplifiers used for SSB.

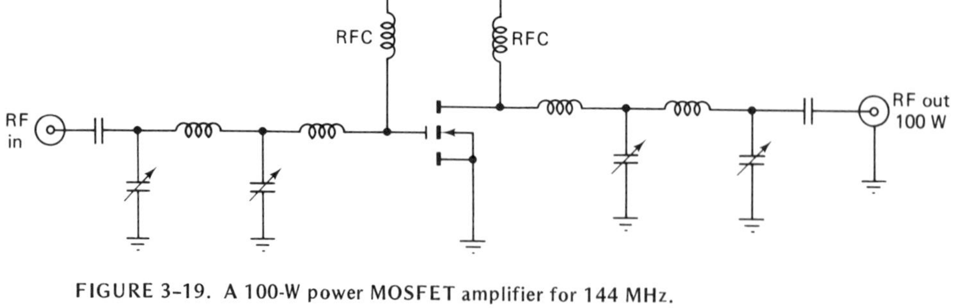

FIGURE 3-19. A 100-W power MOSFET amplifier for 144 MHz.

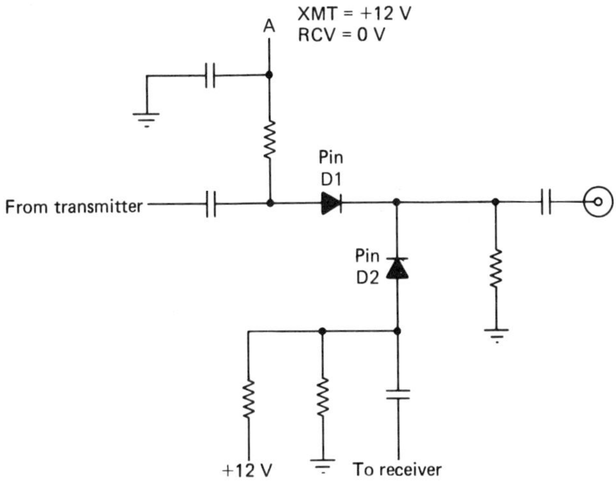

FIGURE 3-20. A PIN diode switch is used in place of the antenna changeover relay.

3.3.6 Diode T/R Switches

One device unique to VHF transmitters is the solid-state PIN diode transmit/receive switch. The RF resistance of a PIN diode depends on the dc current being passed through the device. Therefore, if a heavy forward current is passed through the PIN diode, the diode has a low resistance and is said to be "on." If the diode is reversed-biased so that no current passes through the deivce, the diode is said to be "off." These "on" and "off" characteristics are not valid at lower frequencies, so that the PIN diode switch is reliable only at VHF and higher frequencies. Figure 3-20 shows a PIN diode switch for use in a low-power two-meter FM transceiver. When the voltage at A is positive 12 V, D1 is forward-biased and D2 is reverse-biased, and the transmitter power is coupled to the antenna, while the receiver is protected from damage by D2. when the voltage at A is zero, D1 is "off" and D2 "on," thus connecting the antenna to the receiver while disconnecting the transmitter from the receiver.

PIN diode switches have some slight loss over a relay, but increase in reliability is an excellent trade-off for the slight loss in received signal or loss of transmitter power.

4 SPECIALIZED FORMS OF COMMUNICATIONS

4.1. RADIOTELETYPE

What are classified as specialized forms of communications today may be standard in a decade or so. As technology advances, systems considered difficult or expensive can become commonplace. A good example is that in the late 1940s, single sideband was a specialized form of communication used only by a few amateurs. Of course, single sideband today is the most used emission on the high-frequency bands. What types of specialized communications will exist in the future is a matter of speculation.

The oldest and still popular form of specialized communication is radioteletype (RTTY). Although radioteletype techniques were known for many years, it was not until after World War II that surplus teleprinters were appearing on the market for attractive prices. Since very little commercial equipment was available for sending or receiving radioteletype suitable for amateur use, most avid RTTYers were competent homebuilders. Most of the activity centered around the lower-frequency bands, where it was easy to achieve a stability greater than that required for CW for RTTY. The higher-frequency bands, which would have provided for interesting DX work, were seldom used and very few foreign stations were equipped for RTTY operation anyhow.

This situation did not last for very long. Soon more machines were available and commercial equipment was beginning to appear for amateur use, mainly terminal units used for reception. Foreign stations were becoming equipped for RTTY and the DX attracted amateurs to the higher-frequency bands.

Several good books were published and explained, in great detail, how to connect a teleprinter to an amateur transmitter and how to construct the necessary equipment for a complete RTTY station. Activity was expanded to the 15- and 20-m bands, with the activity on 40 and especially 80 m remaining heavy.

The method of transmitting the teleprinter information is an early and simple form of serial digital data transmission. Long before digital computers were transmitting their information over wires and bits and bytes were household words, personal messages, weather, news, and even tactical information from the battlefield were handled in digital form by the teleprinter.

4.1.1 Teleprinter Codes

A binary code of 5 bits, or elements, is used to transmit teleprinter information. Binary means that the bits are either "0" or "1." Therefore, a 5-bit binary number can take only 32 different forms from 00000 to 11111, which certainly would be a problem if there are 26 letters, 10 decimal digits, punctuation, and the required control functions, such as line feed and carriage return. The 5-bit code is used in the following fashion. First, the teletypewriter keyboard is divided into upper and lower case, much like a regular office typewriter. However, unlike the office typewriter, the upper cases of the letters are not capital letters but numbers or punctuation. Therefore, teleprinter messages contain only upper case. Two of the possible 32 binary codes must be used to place the printer in either upper or lower case. If a "figures" character is sent, all data sent after the "figures" will print the upper case until a "letters" character is sent, whereupon only letters will be printed.

There are three other nonprinting functions: carriage return and space, which are self-explanatory, and line feed, which advances the paper one line. There is one key on the keyboard which has no designator printed on it and is called, appropriately, "blank." This key causes the teleprinter to do nothing and is used for spacing out tapes and testing. The remaining 26 codes of the possible 32 are the letters of the English alphabet.

The upper case of each letter is a number, or punctuation, but not all teleprinters have the same number or punctuation above each letter.

The numbers are usually found at the same place on any teleprinter, but not so with the punctuation. There are several different types of keyboards, which differ only in punctuation. Teleprinters of differing keyboards do not send unusual codes but only have different keytops and type bars installed. If two teleprinters with differing keyboards were communicating, errors would be detected only if the special characters were sent. For most English-language texts the two different machines would communicate correctly.

The digital data representing the 5-bit word designating the teleprinter character to be printed are sent in serial format. That is, each bit has a specific time slot, and if the pulse is present, the data bit is a 1 and if the pulse is not present the data bit is 0. In the radio communications field, the 1s and 0s of the computer world are referred to as marks and spaces. The transmitting and receiving teleprinters are synchronized in time by adding a start-and-stop bit to the digital serial data. Thus, the total number of pulses used to transmit the teleprinter code is seven: a start bit, five data bits, and a stop bit. The teleprinter code, as diagrammed in Figure 4–1, shows time along the horizontal axis while the base line represents the space state of the system and the high of the vertical axis represents the mark of the system. Every pulse is 22 milliseconds (ms) wide except the stop pulse, which is 31 ms wide. The state of the start pulse is always a space, and the stop pulse is always a mark. The five data pulses between the start and stop pulses are either mark or space and depend on which letter is being sent.

Not all teleprinters use the code just described and there is more than one alternative. One of the simplest variations is to shorten the stop pulse from 31 ms to 22 ms, making it the same length as all the other pulses in the system as shown in Figure 4–2. This reduces the time required to send a character from 161 ms to 154 ms and thereby increases the data rate from about 60 words per minute (wpm) to 65 wpm. Further increase in speed to 100 wpm is accomplished by reduc-

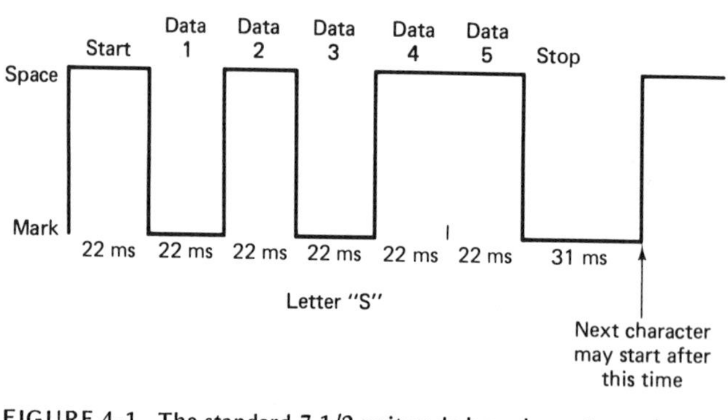

FIGURE 4–1. The standard 7 1/2-unit code has a long stop pulse.

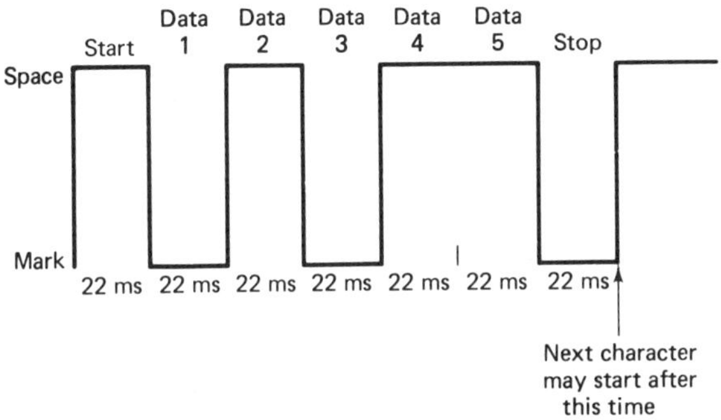

FIGURE 4-2. The 7-unit code has a stop bit that is the same time as the data bits.

ing the pulse width to 14 ms, including the stop pulse. These variations change only the speed and do not affect the number or kind of characters. For many years, the only teleprinter speed permitted on the amateur bands was 60 wpm. Recent rulings by the FCC has allowed 100-wpm teletype and another serial digital data code called ASCII. This code, which is an abbreviation for American Standard Code for Information Interchange, is not really a communications code but a code for interfacing with a computer. However, there are many ASCII teletypewriters that can communicate just as effectively as any conventional teleprinter. The ASCII code has seven data bits which implies that there are 128 different codes, so many that the entire alphabet, upper and lower case, numbers, punctuation, and many special control functions are transmitted. Being a computer code, an ASCII system can be used for direct interface between a computer and a radio system or a computer terminal could be used for receiving conventional messages. Using the ASCII system for radio message transmission would be an advantage since fewer shift commands are required, and there are no variations in keyboards.

There are several ways to transmit the digital signals of the teleprinter. One of the first attempts was to use a simple make–break keying, such as that used for Morse code. However, the teleprinter is very susceptible to noise and even the slightest disturbance will cause the machine to misprint. To gain more noise immunity, the frequency-modulated equivalent of make–break keying, that is, frequency shift keying, was employed with great success. In this method of transmission, a carrier is always present, while its frequency shifts between two distinct frequencies. Limiters and discriminators, as found in any common FM receiver, are employed in demodulation to give the same kind of noise-reduction limiting provided in an FM receiver. The amount of frequency shift chosen at the time FSK was first used was

850 Hz, a value that many commercial stations were using. Although this shift was standard for amateur RTTY for many years, it is much wider than theoretically necessary and was narrowed to 170 Hz when the quality of amateur equipment improved.

Frequency-shift systems require a very stable receiver and transmitter for successful communications. For narrow shift, a stability of 50 Hz during a transmission is required. Modern single-sideband equipment usually has this kind of stability. However, older transmitters, especially VHF types, could not begin to approach this order of stability. Therefore, alternative methods of transmitting RTTY were developed. One system used for teleprinter transmissions, especially on the VHF bands, was audio-frequency shift keying or AFSK. In this system an audio oscillator capable of shifting between two tones was used to modulate a standard AM transmitter. The tones could be practically any frequency in the voice-frequency range but were chosen to be the same as those used for terminal units. This allowed a standard FSK receiving system to demodulate the AFSK signal. The major disadvantage of AFSK is the waste of transmitter power and spectrum space. An AFSK signal has a carrier and two sidebands to do a job that needs only a carrier. Since almost all voice communications on the VHF bands have changed to FM, AFSK using FM has been used with great success. Although not as spectrum-efficient as FSK, FM-AFSK is superior to the older AM version.

4.1.2 Teleprinters

The teleprinter is probably the only piece of amateur equipment that is almost never purchased new. Although manufacturers of teletypewriters will gladly sell a new machine to an amateur, the cost of such a machine is beyond all but the very rich. However, teletypewriters are in very widespread use and abound in the used-equipment markets. Teletypewriters were manufactured with high-quality parts and rugged construction to withstand the rigors of continuous-duty operation, and it is not unusual to see a teleprinter approaching 40 years old still being used in amateur service. The Teletype Corporation Model 15 is an example of a very long life machine. These machines were built for many years both as stand-alone units and as a part of a complete communications set called the Model 19. A later model, and generally superior machine, was the Model 28, which was also made for very rough duty. A much later model, which is a small and attractive machine but intended only for intermittent duty, is the Model 33, which is suitable for amateur use. Other manufacturers, such as Mite

The HAL ST-6000 demodulator features an oscilloscope tuning-aid and other valuable features. (Courtesy of HAL Communications Corp.)

and Kleinschmidt, produce similar machines that can be purchased used at attractive prices.

The usual teletypewriter is called a *page printer* and is only one of the possible machines that can receive teletype messages. Another device used for the reception of teletype signals is the *reperforator*. This machine punches paper tape from teletype signals and is to be differentiated from the paper-tape punch, which can only punch tapes from a keyboard and not from received signals or, as it is called in the wire services, "on-line" signals. Some reperforators can type and punch the received message directly on the paper tape and are called, appropriately, *typing reperforators*. A lesser used machine on the amateur bands is the *strip printer*. This device is similar to a typing reperforator, does not punch holes in the tape, but types messages on a long strip of paper. The reperforator and strip printer occasionally had keyboards that would allow messages to be originated. Punched paper tape can be transmitted from a device called a *transmitting distributor*. The operation and internal appearance is similar to an automobile distributor, hence the name. More modern devices are called *paper-tape readers* and usually use an electrooptical device for reading the paper tapes.

Before purchasing any teletypewriter, it is important to know how to determine if the machine has a standard 5-bit Baudot code and at what speed it operates. Some machines were capable of operating at 100 wpm just by changing a gear lever, while other machines could be internally modified by removing one gear and replacing it with another, and some older machines could not operate at all at 100 wpm. When purchasing a teletypewriter and 100-wpm operation is contemplated, be sure the machine is capable of operating at that speed. In general, the Model 15 could operate only at 60 wpm while most other machines were capable of 60, 75, or 100 wpm, by either changing gears or selecting the speed with the transmission.

4.1.3 Cathode-Ray-Tube Terminals

Another device for sending and receiving teletype messages is the cathode-ray-tube (CRT) terminal. This device resembles a computer terminal except that it communicates with the Baudot code. Any computer terminal, if properly programmed, can be used for radioteletype communications. A brand new terminal can be purchased new for about the same price as a good-quality used teleprinter. The terminal is quieter and can operate at any speed without gear changes. On the disadvantage side is the lack of hard copy. Many of the tasks handled by amateur RTTY require a permanent record, such as traffic handling, satellite schedules, and bulletins transmitted by the headquarters station of the American Radio Relay League. Another disadvantage is the lack of tape punching and reading equipment. Tapes are handy for storing messages, such as CQ calls, descriptions of station equipment, and RTTY art. Some terminals have storage that will allow messages to be stored as long as the power is applied to the terminal. Long-term storage can be done on audio-cassette tapes using the necessary interface equipment.

4.1.4 Teletype Stations

Figure 4-3 shows a block diagram of a complete RTTY station using an SSB transceiver and a conventional teleprinter. This station consists of the minimum equipment required to adapt an SSB transceiver to RTTY. The teleprinter keyboard output keys the AFSK oscillator, which is fed into the microphone input of the transceiver. Many transceivers have higher-level inputs for phone patches or specifically for RTTY equipment. If such an input is not available, the microphone input may be used, but great care must be taken to prevent any ac hum pickup. Audio output from the transceiver is fed to the terminal unit, which demodulates the RTTY signal. A loop supply is shown, but in many cases, the loop supply is internal to the terminal unit.

The loop, or local loop, is used to drive the magnets of the teleprinters, tape reperforators, or other printers. The selector magnet of a teleprinter has quite a bit in inductance and requires a high voltage to ensure that it will operate quickly. A supply of at least 200 V with a current-limiting resistor is required. Usually, all the receiving teleprinters are placed in series to form a large "loop," as shown in Figure 4-4. The switching transistor breaks the current to all the selector magnets placed in the loop and all machines print the same. The dc resistance of the magnets is low compared to the series resistor, so that machines can be inserted and deleted from the loop without readjusting the limiting resistor.

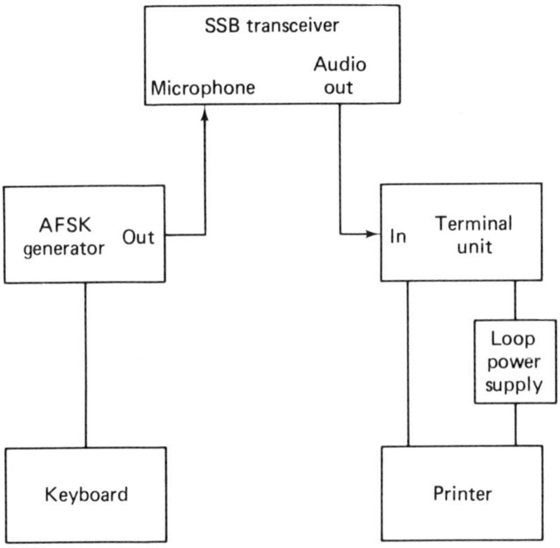

FIGURE 4-3. Block diagram of a typical RTTY station using a single-sideband transceiver.

Power for the magnets is derived from the loop supply and is usually set to either 20 or 60 mA, depending on the requirements of the machine.

On the transmitting side, the keyboards and transmitting distributors are also placed in a loop as shown in Figure 4-5. In this case, no power supply is required and the breaking of the circuit can be accomplished by any key contact. Keyboards and transmitting distributors can be removed from the loop as long as the point in the loop where the device was located is shorted together.

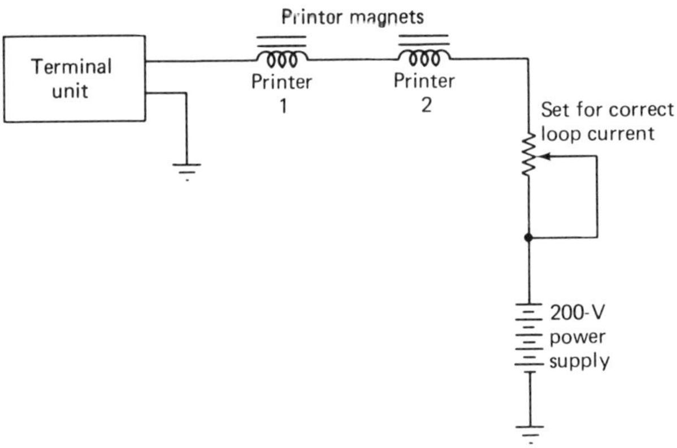

FIGURE 4-4. Receiving local loop, showing two printers connected.

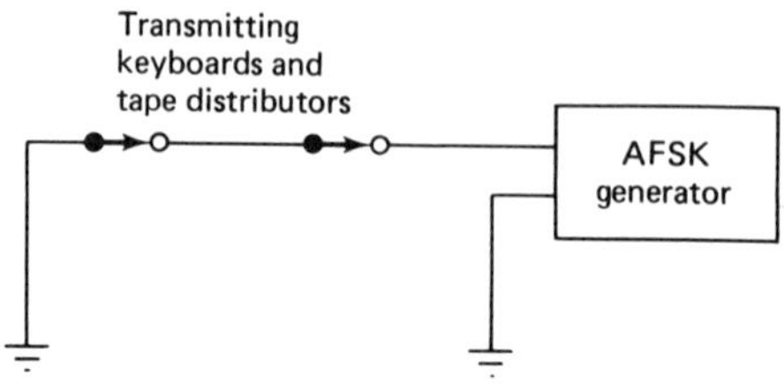

FIGURE 4-5. The transmitting local loop contains only keyboards and distributors.

The connections required for a CRT display and a solid-state keyboard are somewhat different from the connections required for the conventional teleprinter. First, no loop supply is required to drive the printer magnets. The CRT requires only a low-voltage signal, usually in the form of a switchlike closure. A terminal unit is still required to convert the audio-frequency tones to pulse signals for the CRT and to accept the pulse signals from the solid-state keyboard and convert them to audio-frequency tones suitable for feeding the SSB transceiver. Figure 4-6 shows the connections for the HAL Model DS3100 video display terminal.

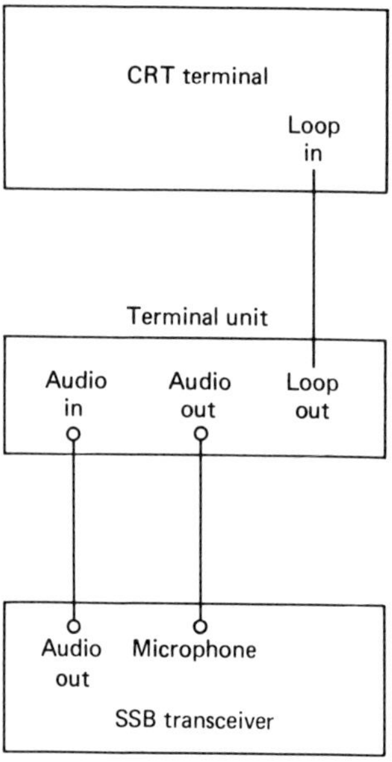

FIGURE 4-6. Interconnections between a SSB transceiver and a terminal unit feeding a video terminal.

The performance of the RTTY station is determined by the design of the terminal unit more than any other piece of equipment. The terminal unit accepts the audio tones from the receiver, processes them, and produces the current impulses required by the teleprinter or CRT display.

4.1.5 Terminal Units

There are two basic types of terminal units available for RTTY use: the limiting type and the limiterless terminal. Often, both types of operation are available from one terminal unit selected by a switch. It has been generally agreed that the limiterless system of demodulation is superior to the limiting type, particularly with weak signals or in the presence of interference. The limiting system is superior when conditions are good and there are few interfering signals.

The block diagram of the terminal unit shown in Figure 4-7 is typical of an amateur terminal and is applicable to both the limiting type and the limiterless type.

The audio input signals are supplied from the receiver and passed through a bandpass filter to remove interfering signals and reduce the amount of noise energy present. The total bandwidth of a 170-Hz, 100-wpm radioteletype signal is about 200 Hz, which is an optimum bandwidth for the filter. A much wider filter is required for 850-Hz shift and a switching arrangement must be included for terminals operating with both shifts. For the limiting system, a limiter follows the bandpass filter. For the limiterless type of system, a linear amplifier immediately follows the bandpass filter.

The use of a limiter produces characteristics similar to those found in an FM radio, most notably the capture effect. The strongest signal presented to the input of a limiter captures the limiter. That means that the output is unaffected by interfering signals when a signal has captured the limiter. If any other signal becomes stronger than the desired signal, even temporarily, it will capture the limiter. On the positive side, the output from the limiter is constant for a wide variation in input levels. This similarity to an FM receiver has been responsible for the limiting type of terminal being called the FM system, while the non-limiting type has been called the AM system.

The output of the limiter, or linear amplifier, drives the frequency discriminator. In its simple form, this circuit consists of two tuned circuits tuned to the mark and space frequencies. One tuned circuit feeds a positive polarity rectifier, while the other tuned circuit feeds a negative polarity rectifier. The outputs of these rectifiers are added together so that if one tone is being received, the output voltage is posi-

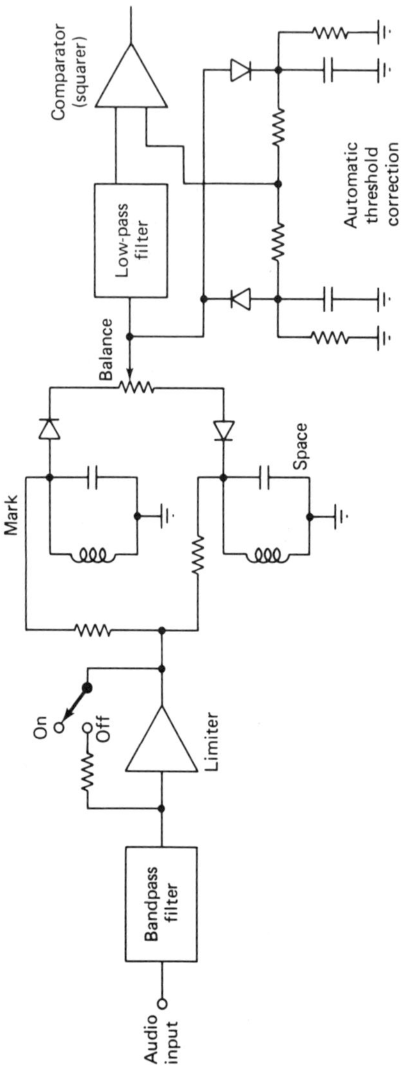

FIGURE 4-7. Block diagram of a basic terminal unit.

tive, and if the opposite tone is being received, the output voltage is negative. The transition from mark to space should be smooth without any sudden changes in voltage. Figure 4-8 shows the shape of the output voltage as a function of input frequency for a well-designed discriminator. The discriminator drives the automatic threshold corrector and the squaring circuit.

The squarer is used to provide a fast rise-time pulse with a well-defined threshold from the discriminator output. This circuit is nothing more than an analog comparator.

When using the limiterless type of circuit, the signal presented to the frequency discriminator is not a constant but will vary with the signal strength. For maximum performance the squaring circuit should receive a threshold that has been corrected for variations in signal strength. The automatic threshold corrector is fed from the discriminator and charges two capacitors to the peak of the discriminator output. One capacitor is charged to the value of the peak mark output and the other capacitor is charged to the peak of the space output. A resistor divider generates a voltage that is precisely midway between the mark and the space peak values. The slicer changes state at this point. If fading should cause the strength of the space or mark signal to decrease, or mistuning should cause one signal to increase while the other decreases, the peak voltage as stored on the capacitors will change and the threshold as determined by the resistor divider will correct for the new value. The time constant of the automatic threshold corrector is very slow, so that only the slow fading and frequency drift will be corrected.

The output of the discriminator is fed to a low-pass filter to reduce the noise level. The bandwidth of the low-pass filter is dependent on the speed of the transmission. The common 60-wpm baud rate is 45.45

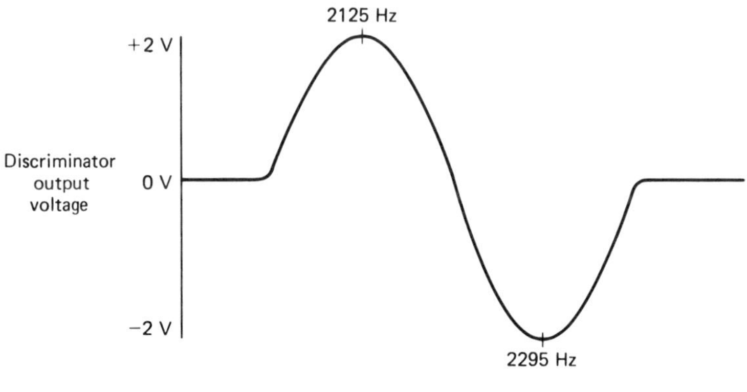

FIGURE 4-8. A good discriminator curve has a smooth transition from mark to space.

Hz, which corresponds to a bandwidth of about 90 Hz. The low-pass filter has a cutoff frequency of about 90 Hz, which allows much of the noise to be reduced without introducing any significant distortion that would cause misprinting.

A valuable accessory for the terminal unit is the auto-start circuit. This system determines that a valid RTTY signal is being received and applies power to the teleprinter motor. This allows the receiver to be left unattended, and the reception of a valid signal will cause the machine to be powered and, after the receipt of a message, turn the motor off. This is not intended to be a power-saving device, but a teleprinter that is operating with noise only will print gibberish and usually ends up piling up at the end of each line. This wastes paper and damages the platten.

One method of implementing auto start is shown in Figure 4-9. This is the method used in the HAL ST5000 and other HAL terminal units. Samples of the mark and space bandpass filter outputs are rectified, unlike the discriminator, both in the same polarity. This signal is called *plus-plus* to distinguish it from the discriminator output, which is *bipolar* or *plus-minus.*

When the signal is present in both the mark and the space channel, the plus-plus voltage is at its maximum. This voltage charges a capacitor and, after a delay, exceeds the threshold of a comparator which applies

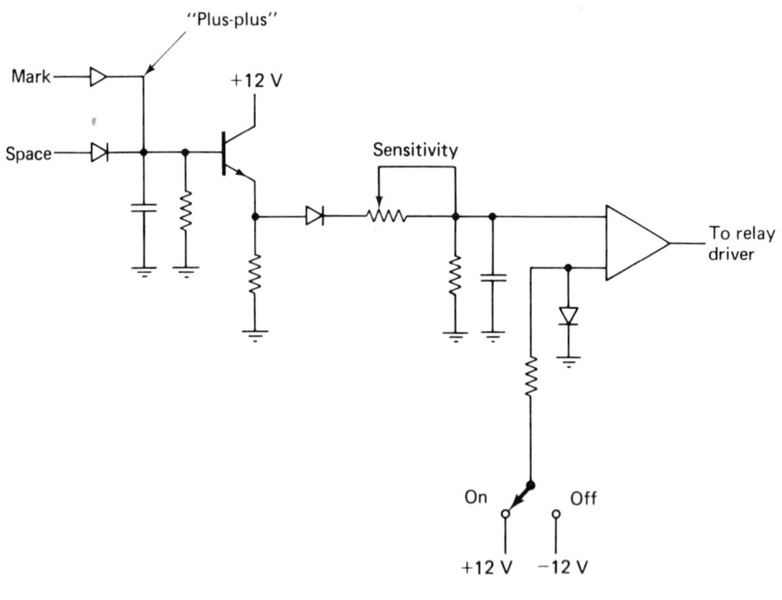

FIGURE 4-9. Auto-start circuit using both mark and space signals.

power to the motor. The cessation of activity in the mark and space channels causes the plus-plus signal to decrease and hence the motor turns off.

4.1.6 Tuning Aids

Before any terminal unit can operate at its peak performance, the RTTY signal must be properly tuned. Tuning the receiver to within 50 Hz or closer is difficult without a good tuning indicator. Since the frequency of the RTTY signal is constantly changing, it is difficult to use a simple meter circuit, and many other schemes have been devised,

Complete RTTY transmitting and receiving facilities are available from the HAL DS-3100 ASR. (Courtesy of HAL Communications Corp.)

mostly using cathode-ray tubes or other electron-beam devices that are fast enough to follow the received data. Even with the most sophisticated tuning device, the most accuracy is obtained with a skilled operator.

The oscilloscope is one of the more popular indicators and provides a reasonably accurate indication of proper tuning. Reception of a mark tone causes a horizontal line or ellipse to he displayed on the oscilloscope screen. Reception of the space tone causes a similar vertical line to be displayed on the oscilloscope. The length of the line is determined by the relative strength of the signal and the position; that is, the angle it makes with the horizontal is determined by the frequency. A perfectly tuned signal will have two lines exactly 90 degrees apart and equal in length. The effects of fading and interference can be readily seen with the oscilloscope.

A tuning meter is a much less expensive tuning device. Two tuning-meter methods exist. The older conventional method is to use the bipolar voltage from the discriminator to deflect a zero-center meter, where proper tuning would be indicated by a centered meter. However, the actual meter position is influenced quite a bit by the data being sent. A series of RYs will produce a steady zero reading of the meter but will begin to drift just as soon as other data are transmitted. Tuning corrections are usually made during a transmission which would not include a series of RYs. A newer method used in the ST5000 terminal unit by HAL uses the plus-plus voltage to drive a left zero meter. Since the plus-plus voltage is maximum when the mark and space frequencies are in the center of the channel, the meter will have maximum deflection.

4.1.7 Radioteletype Transmitters

There are some very important requirements for the transmitter to be used for RTTY transmission. First, and most important, is the duty-cycle capability of the transmitter. RTTY requires that the carrier be present continuously and many transmitters are not capable of operating full power for more than a minute or so, especially SSB transmitters of recent vintage. Some transmitters can be operated continuously at reduced power level and some transmitters cannot be operated continuously at any power level. Very old transmitters that were designed for AM operation can operate at nearly full power continuously and are excellent for radioteletype. Most of these transmitters were the frequency-multiplier type and had poor frequency stability on the higher-frequency bands, but are excellent for RTTY operation on 40 and 80 m.

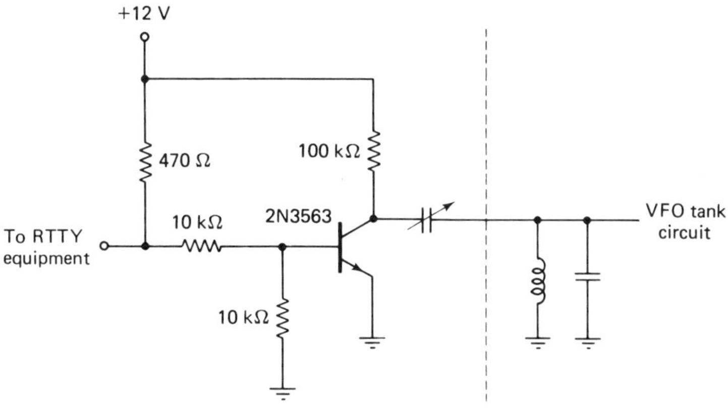

FIGURE 4-10. Adding FSK to an existing VFO using a transistor switch.

A slight modification must be made to an older CW/AM transmitter to be used on RTTY, which requires that an electronically switched capacitor be added to the VFO. Figure 4-10 shows a typical VFO circuit of an older transmitter. In this figure, a small trimmer capacitor has been added to the tank and is switched in by the transistor, which is driven from the keyboard contacts. The amount of frequency shift can be set by the trimmer capacitor. If the old transmitter is to be used on more than one amateur band, the amount of frequency shift will change as the bands are changed due to frequency multiplication. This can be compensated by providing two capacitors, as shown in Figure 4-11. This system causes two capacitors to be switched on the low-frequency band while only one of the capacitors is switched on the higher-frequency band. If this circuit were connected for use on both the 40- and 80-m band, C1 would be adjusted for proper shift on 40 m while C2 would be adjusted without disturbing C1 for the proper shift on 80 m.

Much more current than necessary to switch the transistor is passed through the keyboard contacts. The purpose of this current is to burn away the dirt, oil, and oxides that will accumulate on the keyboard contacts.

If the switched trimmer capacitor is installed in a heterodyning type of transmitter, no adjustment is required to equalize the shift for different amateur bands. Once the shift has been set for any band, it will be correct for all bands.

Single-sideband equipment can be used on RTTY by simply feeding an audio-frequency shift oscillator into the microphone jack. This simple procedure produces a good-quality RTTY signal without any modification to the equipment. However, since linear amplifiers are

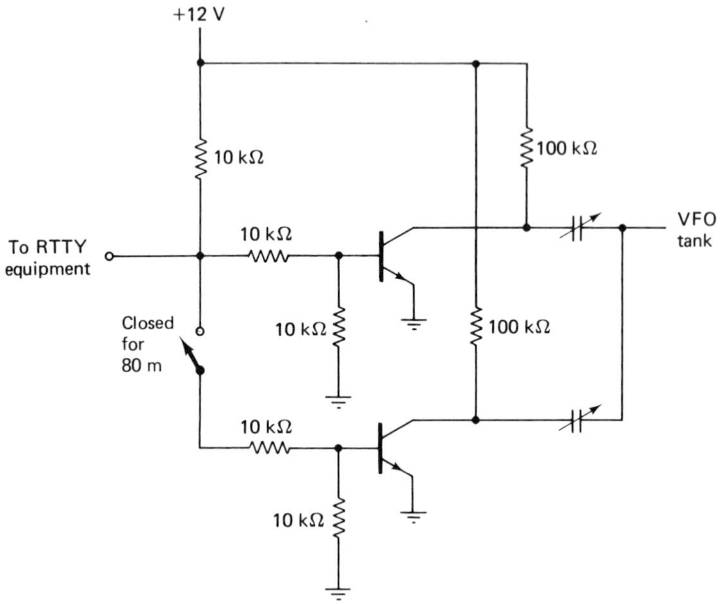

FIGURE 4-11. FSK circuit for a multiplier-type transmitter.

used throughout the transmitter or receiver, and are not efficient amplifiers, the continuous carrier can damage the transmitter if the power is not reduced. The amount of reduction is dependent on the design of the transmitter, and in some cases, can be more than 50 percent.

It is important that the audio-frequency shift oscillator be free of distortion or noise. Any signals, other than the mark or space tone, will appear at the output as a spurious signal. Harmonic distortion will be attenuated by the sideband filter, but ac hum and power supply ripple will fall into the passband and produce strong, spurious signals. An AFSK oscillator using an integrated-circuit sine-wave generator is shown in Figure 4-12. The circuit used to generate the sine wave has low harmonic distortion and can be electronically switched by changing the control voltage. The harmonic distortion is very low, so no further filtering is required after the output. The overall stability of the SSB transmitter will depend to some extent on the stability of the AFSK oscillator, and the stability of the AFSK generator shown is far better than is required for reliable operation.

The final piece of teletype equipment actually has nothing to do with the successful transmission of teleprinter signals. It is the CW identification system used to satisfy an FCC requirement that all RTTY transmissions be identified by International Morse code. The identification may be in the form of normal CW or sent by FSK, which is the

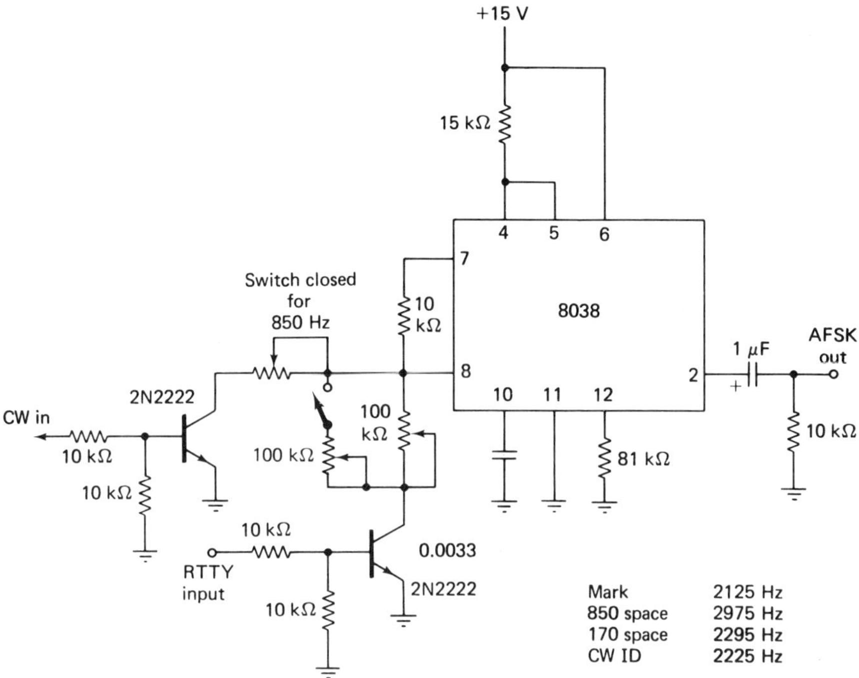

FIGURE 4-12. Audio FSK generator suitable for use with an SSB transmitter for generating FSK.

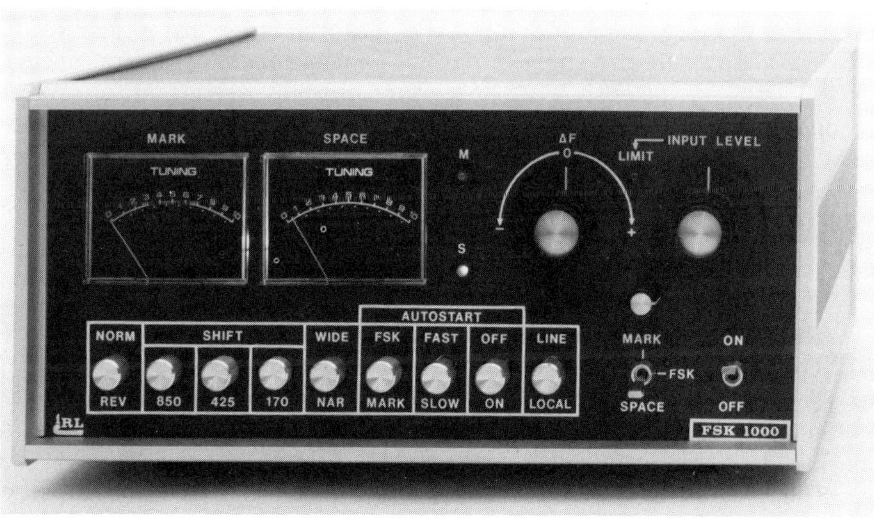

The IRL FSK 1000 terminal unit has a unique system for tuning using 2 meters, one for mark and the other for space. (Courtesy of IRL.)

preferred method. The deviation can be reduced from 170 or 850 Hz, but may not be less than 100 Hz. It is a common practice to reduce the Morse code identification to near the legal minimum so that the teletype terminal will not decode the narrow shift and the teleprinter will not print gibberish. The simplest method of inserting the code is to use a hand key inserted in the normal FSK loop. More sophisticated systems use solid-state digital message generators with the station call stored in a read-only memory.

4.2 SLOW-SCAN TELEVISION

After RTTY, the second significant specialized communication in the high-frequency spectrum is slow-scan television. The nomenclature of television is somewhat misleading, as the pictures are actually stills and cannot impart any sense of motion because of the relatively long period of time required to transmit a complete picture.

Slow-scan television had its beginnings during the 1950s and did not really gain popularity until the late 1960s, when commercial gear became available.

Slow-scan television is capable of transmitting a still picture or slide from either a camera or a slide projector in about 8 seconds. The video information is sent via a single-sideband transmitter on certain portions of the amateur phone bands and is received with a regular communications receiver or single-sideband transceiver, converted, and displayed on a long-persistence cathode-ray tube or stored in a more sophisticated scan storage system. The normal long persistence phosphor decays slowly after the reception of the picture but may be viewed for several seconds after the complete picture has been received. The received picture may also be stored on ordinary audio tape recordings for later viewing or retransmission.

4.2.1 Slow-Scan Specifications

The slow-scan picture consists of 120 lines with an aspect ratio of 1:1, which gives a resolution of 120 lines overall. The picture information is sent from left to right and top to bottom as with standard television. The beginning of each picture and the beginning of each line is marked with synchronizing pulses much like standard television.

The picture and synchronizing information is sent via frequency modulation of a subcarrier. The subcarrier spans from 1200 Hz, which is in the "blacker than black" or infrablack region. A 30-ms pulse is used for the vertical pulse and marks the beginning of each picture, while a 5-ms pulse is used for the horizontal pulse. The video information is sent between the horizontal pulses and determines the frequency of the subcarrier. As dark areas of the picture are scanned, the subcarrier is lower in frequency than when lighter areas are scanned.

The complete slow-scan television picture is sent in about 8 seconds, which corresponds to a line rate of about 15 lines per second. At this slow frame rate (a frame is one complete picture), the total bandwidth required for picture transmission is less than 3 kHz, or somewhat less than a normal voice bandwidth. Any audio information is usually sent before or after the transmission of a picture. However, it is possible and legal to use independent sideband for simultaneous voice and picture information transmission.

Since no amateur transmitter is capable of transmitting two independent sidebands, this technique is never used.

Aside from the obvious, there are some important differences between slow-scan television as used on the amateur bands and the so-called "fast-scan" television used by broadcast stations. First, slow-scan television is transmitted by frequency modulation. The video information is determined by the frequency of the subcarrier and does not depend on the strength of the received signal. This allows the slow-scan picture to operate with signals that are subject to selective fading, noise, and interference. Standard television uses amplitude modulation and is very vulnerable to noise. Second, the slow-scan picture is sent complete from top to bottom in one pass. Normal fast-scan television pictures are sent one-half at a time. During the first pass, only 262 and one-half of the 525 lines are scanned. During the second pass, the other half of the lines are scanned and interlaced between the lines scanned during the first pass. This technique is called interlaced scanning and is done to prevent flicker. The 8-second time required for a slow-scan picture ensures that flicker will not be a problem.

4.2.2 Cameras

Several pieces of requipment, in addition to an operating SSB-equipped amateur station, are required. A camera or flying spot scanner used with a modulator is required for transmission, while a demodulator and a monitor is required for reception. Figure 4–13 shows the connections required for a typical station. Audio from receiver is processed or converted from frequency modulation to amplitude modulation while the synchronizing pulses are separated from the video information. The output of this processor drives a CRT monitor. On the transmitter side, information from a slow-scan camera is converted to audio-frequency tones and is used to modulate an SSB transmitter or transceiver.

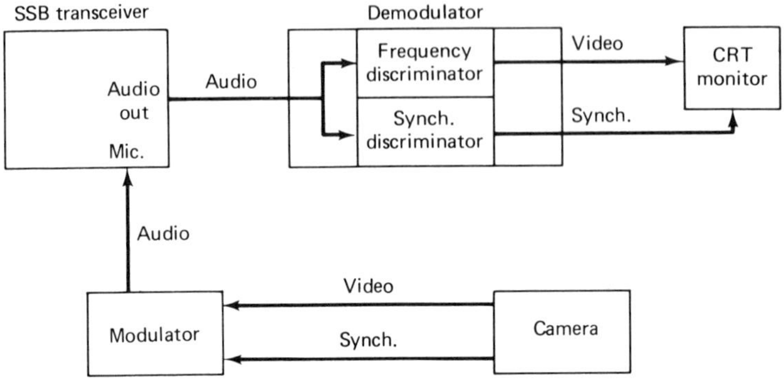

FIGURE 4-13.

There are several slow-scan cameras available to the amateur that use special long-retention vidicons for picture storage. A vidicon is a photosensitive tube that converts the light energy of a scene into electrical signals used for generating television. A block diagram of such a camera is shown in Figure 4-14. This camera uses a shutter much like a film-type still camera that allows the image to be stored on the photosensitive vidicon target. As the electron beam sweeps the target, the image is transferred to minute electrical currents and consequently the image is erased, leaving the vidicon ready for the next frame. To reduce the effects of ac hum and other signals from interfering with the very low level video signals, many of the cameras employ a subcarrier for the video amplifiers. In this camera, the subcarrier is generated by modulating the electron beam of the vidicon, which effectively modulates the subcarrier by the video signal. The video-modulated subcarrier is amplified in relatively narrow-band tuned amplifiers that are unaffected by low-frequency interference such as power supply hum. After amplification, the subcarrier is removed, leaving only the original video information by using synchronous detection.

The camera also provides an electromagnetic sweep for the vidicon and power supplies for focusing and other vidicon requirements.

The video information and the synchronizing information from the sawtooth generators are combined and are used to modulate a voltage-controlled oscillator which feeds a single-sideband transmitter much like an AFSK generator does in generating RTTY.

At the receiving end, a demodulator and a video monitor must be used with a conventional communications receiver. The BFO is used in the receiver and the signal processing is done at audio frequencies, as was done with the RTTY demodulator. Since SSTV is an FM system, the signal from the communications receiver is first passed through a limiting amplifier to remove any amplitude noise. The limiter feeds a

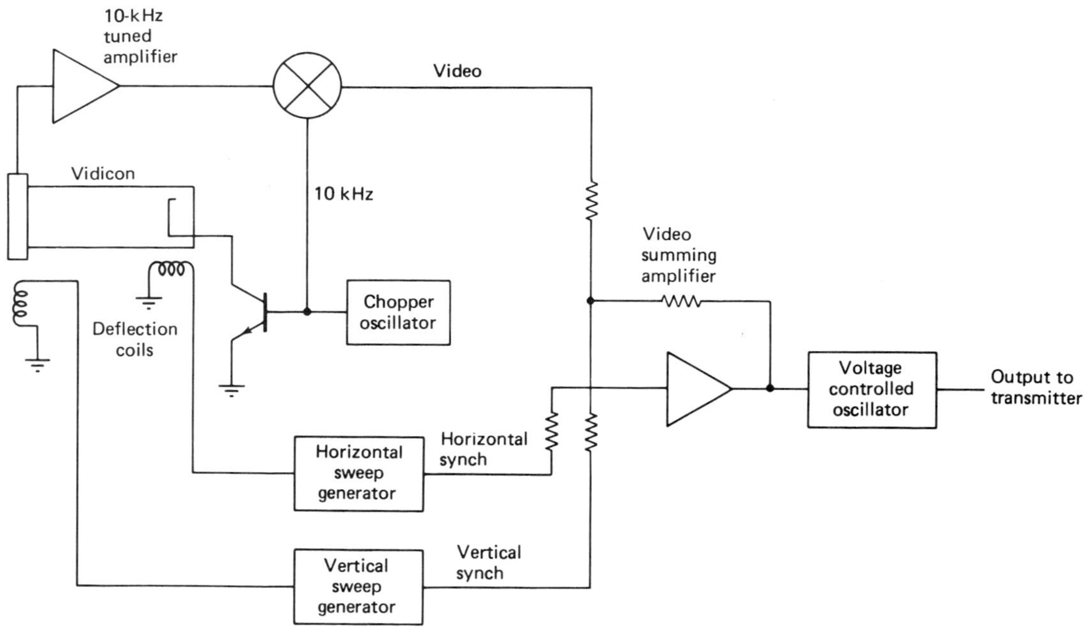

FIGURE 4-14. Block diagram of a typical slow-scan television camera using a subcarrier for noise reduction.

single tuned circuit, tuned to the frequency of the synchronizing pulse, which is used to extract the synchronizing information. Another tuned circuit tuned to 2300 Hz is used as a slope detector for extracting the video information.

Timing circuits are used to separate the long 30-ms vertical pulse from the shorter 5-ms horizontal pulse and the pulses are used to synchronize the horizontal and vertical sweep circuits. The video information is used to modulate the cathode current of the cathode-ray tube, and thus the tube paints a television-like picture on the screen.

The most significant disadvantage of this simple slow-scan receiving system is the relatively short time the picture is viewable. The phosphor begins to decay as soon as it is excited and after a few seconds, the picture is not visible. Even when the picture is fresh, it is best viewed in a darkened room. Several methods have been used to enhance the visibility and the viewing time of the slow-scan picture. One utilizes a special storage tube used in expensive instrumentation. These tubes are not readily available to the amateur and are rather expensive to be purchased new. The storage time on these special tubes is not much more than a few minutes; however, they are bright and easy to see. The best method is to digitize the picture and store it in an electronic memory and replay the picture at a rapid rate. This method is called scan conversion since the slow-scan picture is converted to a faster scan, such as the normal 525-line television.

4.2.3 Scan Conversion

The scan conversion technique can also be used to capture a single frame from a conventional television camera, store the picture in memory, and transmit the picture at the slow-scan rate. This allows a conventional television receiver to be used as a slow-scan monitor and a conventional television camera to be used for slow-scan transmission in addition to providing unlimited viewing time.

To understand the digital storage of a slow-scan television picture, the picture must be characterized as a digital array. The picture is considered as an array of dots, much like the microscopic dots that make up a printed picture, rather than a collection of scanning lines. Since the slow-scan picture consists of an aspect ratio of 1:1 and the picture consists of 128 lines, each line has the equivalent of 128 dots. Each dot consists of one picture element and is called a "pixel." This dot determines the brightness of the picture at that particular location, and sending the value of each dot, one by one, will eventually create the entire picture. The value of brightness varies from black through several shades of gray to white. Sixteen levels of gray, from black to full white, are sufficient to produce good-quality pictures. Sixteen levels or states represent four binary bits, which implies that 4 bits of data must be stored for each pixel. The total information required to store an entire picture is equal to the total number of pixels, 128 × 128 times the 4 bits required for each pixel, which yields a total of 65,536 bits. This is not an unusually large memory requirement and can be implemented with four 8 × 2048 memories, which is a popular item for microprocessors.

Figure 4–15 shows the block diagram of the Robot Model 400 slow-scan television scan converter. This unit is capable of converting the output of a conventional television camera to slow scan and converting slow-scan pictures to fast scan. This is accomplished by digitizing a television picture, either fast scan or slow scan, and storing the picture in a memory. Once the digitized picture has been stored, it may be replayed for any length of time and at any speed.

Referring to the block diagram, the input signal is conditioned, which would be a complete demodulator for the slow scan and a simple level adjustment for the fast-scan input. Both fast scan and the demodulated slow-scan video are digitized with a parallel 4-bit analog-to-digital (A/D) converter. The 4-bit converter can supply 16 shades of gray, which is sufficient for good-quality pictures. The requirements for digitizing the slow-scan video are easily met, but the digitizing of the fast-scan video requires a very fast converter. This is accomplished by using a parallel converter which digitizes the input signal into Gray code numbers. It is a coincidence that the shades of gray are digitized into numbers of the Gray code. The capitalized Gray refers to a code invented by the man whose name it bears, and the noncapitalized gray is

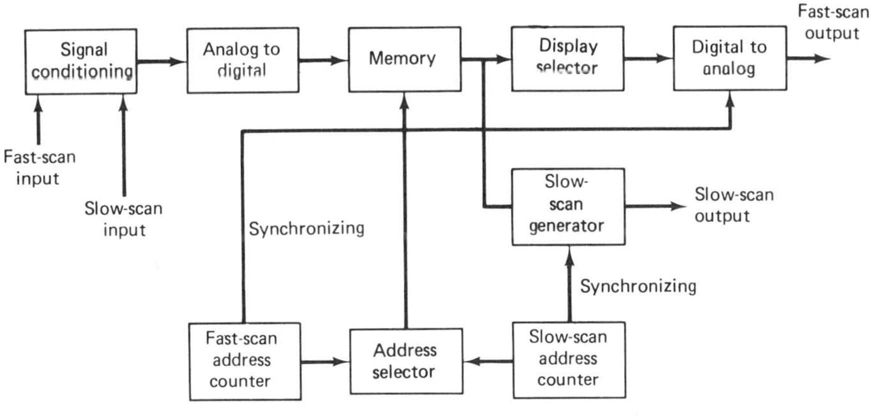

FIGURE 4–15. Scan storage unit capable of converting from fast scan to slow scan and slow scan to fast scan.

the shade of gray in the picture. The unique characteristic of the Gray code is that only one bit changes in transition between any consecutive numbers. The Gray code is used whenever there is a real possibility of a mistake being made in the interpretation of a binary number. If only one bit changes between any two consecutive numbers and that bit were misinterpreted because of the change, the resultant number would only be in error by one. If a binary number were misinterpreted due to changing bits, the error could be very large because it is possible that all the bits were changing. The Gray code outputs are stored in the memory until all the memory locations are filled.

The memory is addressed from two possible counters, called the fast-scan counter and the slow-scan counter. If a fast-scan picture is being captured, the fast-scan counter driven from the fast-scan clock addresses the memory. A digital multiplexer is used to switch the address inputs of the memory from one address counter to the other. For slow-scan picture storage or replay, the slow-scan counters are electronically selected. The fast- and slow-scan counters also supply the synchronizing pulses required for either mode of operation.

When a picture is being viewed, the digital words stored in the memory must be converted back to analog values, and this is accomplished with an analog-to-digital converter. The converter is unique because it is required to convert Gray code to analog voltages, whereas most D/A converters convert binary code. Since the output can be either slow scan or fast scan, the D/A converter, like the A/D converter, must be fast.

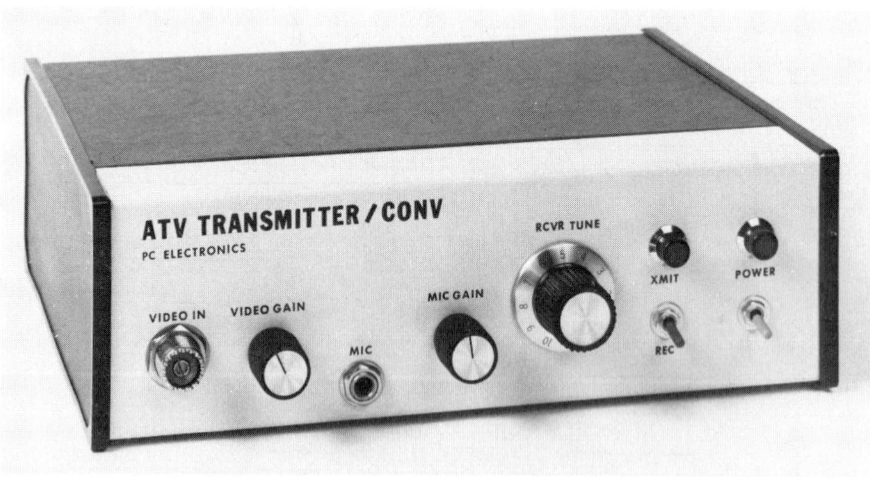

The P.C. Electronics television transmitter/converter is one of the few commercial amateur television products. (Courtesy of P.C. Electronics.)

The Robot scan converter has some circuits incorporated to aid in setting up the proper contrast levels for the converter. A gray-scale generator is supplied which is used to set the gains of the entire system by capturing a test pattern in the memory and replaying the test pattern while the systems gains are set correctly. It is important to know that the picture is stored in the memory with the proper gray scale, since only 16 shades of gray are available. If the entire memory capacity is not used, some of the gray scale will be lost. If the system gains are set too high, too much contrast will result in a poor picture.

An ordinary single-sideband transmitter is required for transmitting slow-scan television, but special care must be taken to ensure sufficient cooling and derating of the transmitter used. The same duty-cycle restrictions for radioteletype apply to the transmission of slow-scan television. Many older single-sideband transmitters or transceivers did not specify duty-cycle or power capabilities while operating slow-scan or teletype. Since the units were rarely, if ever, used for the transmission of these modes, no claims for the performance were made. Most modern equipment that claims to operate full power for slow scan or teletype usually requires the installation of a fan or blower. Even with the blower, it may be necessary to reduce the power input to a safe level. If the transmitter to be used for slow-scan or radioteletype does not specify operation for these modes, it is best to reduce the power input to at least one-half of the peak envelope power of the single-sideband signal. By keeping an eye on the internal temperatures and the tube life, if may be possible to increase the power in steps before the tube life is significantly reduced. when changing tubes, look carefully for signs of overheated components. This will be a sign that the power must be reduced.

4.3 "FAST-SCAN" TELEVISION

Before the advent of slow-scan television, the only amateur television was based on the broadcast industry standards and was authorized on the 450-MHz band and above. Since there was no other form of television, it was not called fast scan until confusion arose with the appearance of slow-scan television.

Early amateur television operations required an enormous amount of equipment and a high level of electronics competence as well as plenty of patience. There was no supply of equipment specifically designed for the amateur, and very little available for affordable prices. Just about the only source of television equipment was obsolete broadcast equipment. In the very early days of amateur television in the 1950s, surplus television equipment was large, cumbersome, and in some cases, was too primitive even for amateur use. Closed-circuit cameras soon appeared and the most important piece of equipment became available and, in many cases, affordable. The television operator usually chose to build the required distribution and switching equipment and to modify a surplus UHF transmitter for transmission.

4.3.1 Station Requirements

Fast-scan television is authorized on all amateur bands above 420 MHz, with most activity being on that band. Television is definitely not a distance (DX) mode, for several reasons. First, there is absolutely no DX propagation on the 450-MHz band. This is not true of CW or SSB operation, where high power and sensitive receivers can be used to take advantage of ducting and scattering in achieving DX. Because of the wide-band nature of the television signal, high signal levels are required to overcome the noise present in the very wide receiver bandwidth. A signal level of roughly 50 microvolts (μV) is required for a viewable signal.

Therefore, most television contacts are with local amateurs and are prearranged by telephone or from a contact on another band. Recently, television repeaters have opened up interesting possibilities for moderate distances on television. The line-of-sight distance from a mountaintop can easily be 50 to 60 miles, which will allow two-way distances of more than 100 miles. Metropolitan-area repeaters located on high buildings can bring together a large number of television-equipped amateurs on a common frequency.

Television is definitely a mode for experimenters, even after thirty or more years of amateur involvement. The equipment requirements are not as great as they once used to be and the availability of cameras and other video equipment is much greater. Still, a high level of expertise is required and a certain amount of home-constructed equipment is required. A successful amateur television station is the mark of a technically knowledgeable amateur.

The wide-band nature of the television signal is the cause of numerous problems. An amateur television signal can be anywhere from 2 to 5 MHz wide if vestigal sideband emission is used and twice that value if both sidebands are transmitted. Vestigial sideband is an amplitude-modulated signal with a full carrier but with only one sideband. This extreme bandwidth requirement is the reason that television is authorized only on the UHF spectrum. For example, a typical television signal would occupy the entire 2-m band. Transmitters, modulators, and antennas must be designed specially for television operation if reasonable-quality pictures are to be obtained.

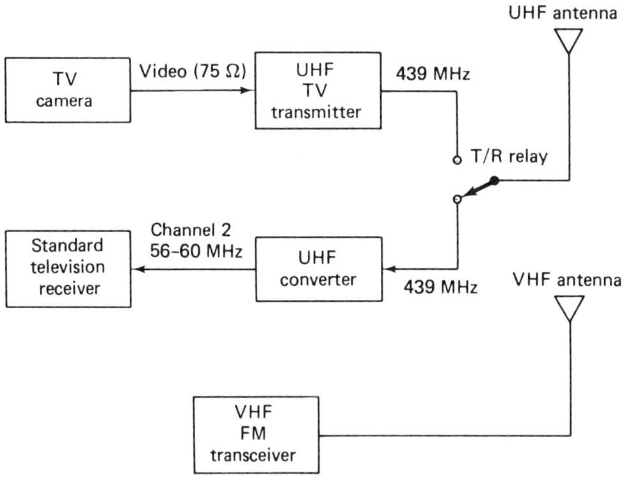

FIGURE 4–16. Amateur television station using a camera and a conventional television receiver.

Narrow beamwidth high-gain antennas are usually narrow band. That is, the SWR is low over only a few megahertz at 450 MHz. When using a narrow-beamwidth antenna, especially with a television repeater, the antenna will not be a good impedance match over the entire frequency range required. It may be desirable to use separate transmitting and receiving antennas when using television repeaters.

A block diagram of a television station is shown in Figure 4–16. This represents the basic equipment required for two-way television communications, with the audio portion of the transmission on a separate amateur band. Commercial television broadcasts the sound portion of the television signal on a subcarrier which is removed from the main picture carrier by 4.5 MHz. The usual method for generating the subcarrier is to use a separate UHF transmitter for the audio carrier and to combine the two transmitters with a duplexer. A duplexer is a high-power filter that prevents the power from the aural transmitter from entering the visual transmitter, and vice versa. This is a very reliable method but requires two UHF transmitters and an expensive duplexer. It is for this reason that amateurs usually transmit the audio with an existing facility, such as 2-m FM.

Another difference between the amateur television transmission and the standard broadcast standard is the amateur use of double sideband. Commercial television stations are required to remove one of the two sidebands and transmit the full carrier and only the upper sideband. This is usually accomplished with a sideband filter applied to the output of the video transmitter, which diverts the unwanted sideband power into a dummy load where the energy is transformed into heat. Other techniques may be used where single sideband is generated and amplified in linear amplifier stages. For amateur use, a common television receiver is capable of receiving a television signal that has both sidebands and does not have a sound subcarrier.

The television receiver is used for an IF amplifier and demodulator being fed from a high-performance 450-MHz converter. Some amateurs have successfully modified UHF television tuners to cover the 450-MHz amateur bands, but this is not desirable because of the lack of an RF amplifier in the typical television tuner. A modified tuner could be used successfully with a preamplifier, however.

4.3.2 Television Transmitters

Television transmitters are usually modified commercial units. A land mobile band, slightly higher in frequency than the 450-MHz amateur band, is a good source of UHF transmitters for modifying for amateur television use. Transmitters used in this band are frequency-modulated and produce power outputs in the range 5 to 50 W. Modification techniques are relatively simple. First, the operating frequency must be lowered to the amateur band. This can be accomplished by adding capacitors to the tuned circuits and supplying a crystal for the amateur band. The second simple modification is the disabling of the FM modulation, which can most easily be done by just turning the deviation control all the way down and removing the microphone. In most commercial FM transmitters, the bandwidth of the power amplifier tank will not be sufficiently wide to pass the television signal. The usual method for increasing the bandwidth of the tank is to change the matching of the tank to the output tube or transistor. Most transistor transmitters for UHF are inherently very broadband and do not require any rematching.

Finally, a video modulator must be constructed and added to the transmitter. An FM transmitter usually operates class C, so that the output stage must be modulated. Because of the wide bandwidth involved, low-level modulation is usually used. A high-level modulator that would be capable of modulating several megahertz would be difficult to build. When using the grid modulator, the operating parameters of the final amplifier must be set correctly for a low-distortion video signal. Several articles in amateur publications describe modifications to popular UHF mobile transmitters for television operation. It is recommended that any modifications follow the instructions given in an article unless the test equipment and the expertise is available to troubleshoot a video transmitter.

Aside from the aural (sound) and visual transmitter, video equipment is required. The least amount of video equipment that is required is a camera, as shown in the television station block diagram of Figure 4-16. Television cameras suitable for amateur television are available both as surplus and new. New cameras include a wide selection of closed-circuit cameras used for television security systems and for home video recorders. These cameras usually have fixed-focus lenses and in some cases have random interlace. Random interlace refers to the way the television picture is scanned with the camera. A television picture is sent one-half at a time. The odd lines are scanned first and half of the picture is sent. In the next pass, the even lines are sent and the second half of the picture is transmitted. This method is used to reduce the flicker from the 30-picture-per-second frame rate. In a good-quality TV camera, the odd and even lines are perfectly interlaced, that is, the odd lines fall exactly between the even lines. This is a function of the synchronizing pulses, which must be perfectly phase-locked. This involves deriving the vertical synchronizing pulse from the horizontal synchronizing pulses with frequency dividers. Before the days of integrated circuits, this was an expensive circuit and the method of random interlace was often used to save cost. In the random system, the horizontal and the vertical are not phase-locked and the interlace is hit or miss. For many years, even after integrated circuits were reasonably priced, random interlace synchronizing was still being used. Many

The Robot model 400 scan converter which allows conventional television equipment to be used for slow-scan transmission and reception. (Courtesy of Robot Research Inc.)

closed-circuit cameras have this deficiency and are not suitable for amateur television. Although the camera will operate, the quality of the picture is inferior. In many cases, it is not difficult to include interlace synchronizing by installing a small number of integrated circuits.

4.4 FUTURE MODES

The future for specialized communications can only be estimated from the current regulations and trends. Recent rulings by the FCC have allowed the use of ASCII teletype transmission on most of the high-frequency, VHF, and UHF amateur bands. The new ruling calls for much higher data rates, which will be especially interesting to computer users who have the need to transmit a lot of data. The main intent of the ruling was to allow for high-speed data transmission between computers as well as conventional keyboard-to-keyboard conversations.

Actually, which bands and to what extent ASCII will be used on the amateur bands for computer communications cannot be determined at this early stage of development. There are many amateurs who are computer enthusiasts, so there is a large potential group of users. Most likely activity will center, at first, in the usual RTTY bands and perhaps expand to the higher-frequency bands, including VHF, for more dedicated high-speed data transmission.

Satellite transmission and reception has existed since the first orbiting satellite-carrying amateur radio, dubbed OSCAR, was launched in 1965. Operating the satellite does not require much in the way of specialized equipment, but requires a knowledge of technique. The times of passes must be known, which can be obtained from data obtained from bulletins transmitted by the American Radio Relay League or from published tables. The position of the satellite should be determined and the antenna pointed and tracked in the correct direction. The amateur bands used for the previous satellites have included the 10-, 2-, and 420-MHz bands. Future satellites have planned on continued use of these bands.

The antennas used for satellite communication are not large or expensive. Only moderate gain is required and high effective radiated powers are not encouraged. The only thing that would mark the satellite antenna from the terrestrial antenna is the use of circular polarization. This is a combination of vertical and horizontal polarization which reduces the fading due to the tumbling of the satellite in space. A circularly polarized antenna is easily constructed by using two conventional antennas mounted on a common boom and fed by a transmission-line phase-shift unit. New satellites will certainly increase the activity and type of operation with spacecraft.

Another type of specialized communication of the near future is a medium-scan television with a frame rate fast enough to allow the transmission of some motion. The bandwidth of the medium-scan television will probably be somewhere between the current slow-scan rate and the current fast-scan rate. Experiments are being carried out with a frame rate of two pictures per second, which is 16 times as fast as the current slow-scan rate and 1/15 the rate of the current fast-scan, or broadcast, rate. The corresponding bandwidth of the medium-scan television is on the order of 36 kHz and will require a change of FCC rules and regulations. However, rule changes were required for radioteletype, slow-scan television, satellites, and other currently popular modes of transmission.

Narrow-band voice modulation is a technique of removing portions of the voice spectrum, repositioning the remaining spectrum, and transmitting the modified spectrum with conventional SSB. This results in a reduced bandwidth for the single-sideband signal which can be reconstructed at the receiving end. The distortion of the received signal is not objectionable and does not degrade the signal to make copying difficult. This type of reduced-bandwidth transmission is not compatible with receivers not so equipped. How many receivers and transmitters will be equipped with the conversion units is another item for speculation.

Another type of communication gaining some popularity in more recent years is the use of the microwave bands, especially the use of the 10-GHz band. Several low-priced solid-state oscillators and other microwave components have become available to the amateur for use on the 10-GHz band. These transmitters operate several milliwatts output and the transmitter is used for the receiver local oscillator, resulting in a duplex operation. Distance records are measured in miles and are strictly line of sight. There is absolutely no atmospheric bending or reflection of the 10-GHz wavelength.

The future of specialized communication is the future of amateur radio. If the past is any indication of the future, there are many interesting and exciting possibilities for the radio spectrum.

5 SELECTING A TRANSMITTER, RECEIVER, OR TRANSCEIVER

5.1 DETERMINING THE PRICE RANGE

For many amateurs, most of the equipment used on the air will be purchased rather than built. For some, the purchases will be of new equipment rather than used or older equipment. Selecting the right equipment for the type of operating intended will make the difference between a sound investment or the wasting of valuable cash. Before any investigation of the types of equipment available is made, prospective buyers must decide at what price range they are willing to invest. That does not mean that buyers must know all the prices before they select equipment, but that buyers should know if they wish high-priced, medium-priced, or low-priced equipment. It is difficult to set a price category before all the prospects have been investigated, and it is equally difficult to choose the equipment that will do the best job without knowing, roughly, what price a buyer wants to pay. Therefore, it is suggested that a price range—low, medium, or high—be determined and then the best equipment in that range be selected for the intended type of operation. If you find that the price of desired equipment is more than you want to pay, your choice of price category was too high. If you want top-priced equipment and find you cannot afford it, you have selected the wrong group. It is not a good idea to buy equipment that is too expensive because once you are committed to a certain manufacturer, you are almost obligated to buy all the accessories for the ham shack from that manufacturer if you desire easy interconnection and matching cabinets. Once you have selected a high-priced line of equipment, the expense will continue as long as you maintain that line. On the other hand, if you select a low-priced line, you will not be able to upgrade without discarding all the equipment purchased. It may be a good idea to purchase a high-priced line of equipment and reduce or defer the purchase of accessories in order to have a higher-quality basic station.

Once you have selected a price category, equipment catalogs, price sheets, advertisements and, in general, any data available on equipment fitting your price category should be obtained and studied. Become familiar with the manufacturers and their model numbers, as this will be helpful for assembling the opinions of current owners. The next step is to listen to the ham bands and hear what others have to say about certain types of equipment. One important advantage for amateurs buying equipment is their ability to communicate. You do not have to transmit, just listen; almost every QSO has some reference to the equipment used. If an amateur does not like a certain piece of equipment, sooner or later he or she will tell someone about it. Also listen to the quality of the transmissions and the ability of the receiving station to get a "good copy." In this way, certain qualities of a transmitter can speak for themselves. Going on the air and asking "How do you like your equipment?" will amost certainly bring a complete answer full of details.

Write down these comments and have the ads and catalogs handy to refresh your memory on some of the specifications. By now you should have some idea of the good and bad points of the equipment and the opinions of several amateurs. It is necessary to review the equipment specifications to determine which equipment can actually perform the modes and types of operation required.

First, list all the types of equipment that you operate and would like to try in the future: that is, single sideband, CW, slow-scan television, contests, field day, and so on. Now, determine which equipment can operate in the modes intended. Many pieces of equipment are advertised to operate in a certain mode but when the specifications are investigated carefully, it can be seen that the performance is so badly degraded that the mode of operation is barely feasible. For example, if a transceiver is advertised to operate CW, it should have installed, or provisions for, a CW bandwidth filter. Using the SSB filter for CW does not constitute good CW operation. Similarly, many single-sideband transceivers will operate on RTTY or slow-scan television. However, investigate the power restrictions when operating these modes. It is possible that the power could be degraded to as low as only 25 percent

The Astro 150 by Swan is a small, fully synthesized transceiver for 80 through the 10-m band. (Courtesy of Swan Electronics.)

of the PEP or CW powers. Make sure that all of the accessories required for operating a certain mode are available for the equipment.

Whether to use a separate transmitter and receiver or to use a transceiver was once an important decision. Most modern communications equipment uses transceivers, which makes the decision much easier. There are some exceptions, but these separates are either very expensive units or for a specialized station, such as CW only. Most separate transmitter and receiver pairs by one manufacturer can be interconnected to transceive, that is, the transmitter's frequency can be controlled by the receiver, and vice versa. If a receiver is used with a transmitter of another manufacturer, this transceive ability will be lost. There is a great amount of versatility in separate transmitters and receivers but there is a very distinct advantage in terms of ease of operation in the transceive mode. Therefore, if a separate transmitter-receiver pair is contemplated, it is not recommended that two different manufacturers be selected.

5.2 SOLID-STATE CONSIDERATIONS

Many amateurs have an inherent fear of solid-state circuits. It is understandable that in a fast-moving technology such as electronics, it is difficult to keep up with the latest in circuits and electronic devices.

Very few modern transceivers use vacuum tubes in the receiver and low-power transmitter stages, but a large number of transceivers use vacuum tubes in high-power and transmitter output amplifiers. This is not to imply that transistors are incapable or unreliable in these stages, only that the manufacturer is taking advantage of vacuum tubes where they can still be cost-effective. For powers in excess of 500 W, vacuum tubes are a necessity. Should they become worn out or damaged, they are simple to replace and often readily available at a local electronics distributor. Many amateurs, if they find the power output lacking from their transmitter, will replace the final amplifier tubes and either fix the problem or verify that the trouble is elsewhere. Vacuum tubes can also withstand a certain amount of mistreatment. Operating the transmitter into a large SWR will shorten the life of the tubes but will be spread out over a relatively long period of time. Transistors, when used as final amplifiers, are quite a bit different. First, they are very difficult to replace. Changing the output transistor usually involves a lot of disassembly and soldering. Changing output transistors is not recommended as a troubleshooting technique. When an output transistor is replaced by an experienced technician, he or she is almost certain that the device is bad. Second, operating the transistor into a high SWR can seriously affect the life of the device. In fact, operating into an extreme mismatch can cause an immediate failure. Finally, the price of a transistor of any power level above a couple of watts is more than that for a comparable vacuum tube. Then, why use transistors in the power-

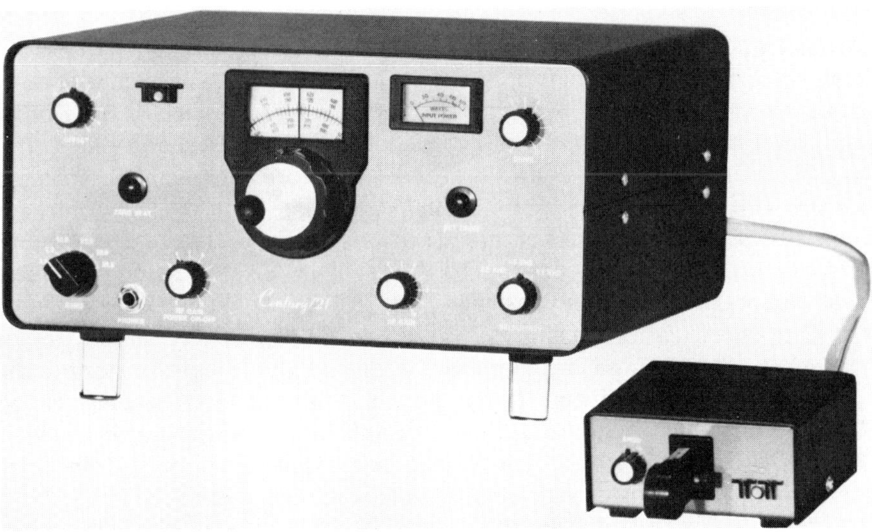

The Ten-Tec Century 21 transceiver is a CW-only unit with special CW-only features. (Courtesy of Ten-Tec Inc.)

output stage of a radio transceiver? If properly designed with adequate cooling and protection from mismatches, the power-output transistors should never have to be replaced during the lifetime of the equipment. It is therefore not the transistors that determine the reliability of the output amplifier, but the peripheral circuits and mechanical design of the transceiver.

Another thing the long-time amateur may fear is the complexity of the equipment. During the early days of radio, most amateurs maintained their own equipment, for several reasons. First, it was less costly to do their own repairs, and except for the factory, there was often no one locally qualified to do the job. Many modern transceivers have become so complicated that very few amateurs can maintain them. This is unfortunate because many of the more complicated designs are good-performing units with impressive specifications. The only possible solution to this problem is a design with special provisions for troubleshooting and an instruction manual that the amateurs can use to repair their own equipment. Another possibility is an exchange system for certain subsystems so that the critical alignment can be done with the proper equipment at the factory. Until a good system is developed, complexity of equipment will be an important consideration when amateurs intend to do their own maintenance.

5.3 SPECIFICATIONS

When comparing the specifications of a transceiver, the only ones that really count are the receiver specifications. Most transmitter specifications are related to the emission of spurious signals, such as harmonics and intermodulation, and are covered by FCC regulations. All transmitters must be type-accepted by the FCC before they can be marketed in the United States, which implies that they will be adequate for amateur transmission. Other specifications, such as frequency stability and dial accuracy, apply equally to the receiver. Therefore, the numbers that can make or break a transceiver are the receiver specifications.

One specification that pertains only to the transmitter and is one of the most often touted numbers but is one of the least important is the transmitter power output. If the power output of a transceiver is too low, it can be easily amplified with an external amplifier up to the legal limit. On the other hand, if a receiver lacks selectivity, or suffers from overload and intermodulation, no external add-on can fix it.

So when buying a transceiver, treat it as if it were a receiver when evaluating the specifications. Therefore, what specifications are important and what conveniences are desired in the transceiver depend mostly on the type of operation the amateur is engaged in. For the purposes of selecting a transceiver, amateur operation will be subdivided into the following classes: CW operators, SSB operators, traffic handlers, RTTY operators, contesters, and DX hounds. Of course, most amateurs fit into more than one category, and some of the requirements for equipment will have to be compromised.

The casual SSB operator will have the widest choice in transceivers. There are useful items, such as passband tuning, digital readout, and VOX operation, but in general, casual operating requires the least from a SSB transceiver.

The CW operator, on the other hand, must have a transceiver capable of operating on CW, which most transceivers manufactured claim to have. If the transceiver uses only the normal SSB filter and has no provision for a narrow CW filter, it does not qualify as a CW transceiver. A 300- or 500-Hz filter is required for serious and casual CW operation.

For the RTTY operator, the usual problems with long periods of "key down" prevail. A transceiver suitable for RTTY will be capable of operating at some power level continuously and the manufacturer will say so. When selecting a transceiver, be sure that the specifications clearly state under what conditions the unit can operate "key down" continuously. In some cases it requires a reduction in drive, and other cases it requires the use of an accessory fan or a fan plus drive reduction.

When selecting a SSB transmitter or transceiver for use on radioteletype, it is important to investigate the level of carrier and unwanted sideband suppression if it is intended to use an AFSK oscillator driving the audio input of the transmitter. As discussed in Chapter 2, modulating a SSB transmitter with a single audio tone produces a single output frequency. If the carrier is not completely suppressed, a spurious signal will appear above the desired signal if the transmitter is operating lower sideband and below the desired signal if the transmitter is operating upper sideband. The amount that the signal will be separated from the desired signal is equal to the frequency of the audio oscillator. If the unwanted sideband is not sufficiently suppressed, another spurious signal will appear above or below the desired signal, again depending on whether the upper or lower sideband is being used, separated by twice the frequency of the audio oscillator. When operating normal SSB in the phone portion of the band, these undesired signals are annoying but legal. Certainly, the carrier and the opposite sideband of an AM signal are legal as long as full carrier double sideband AM is legal. However, on the CW portion of the band, where RTTY would be transmitted, the unwanted sideband and the carrier are spurious signals and are illegal. To comply with FCC rules, these signals should be attenuated at least 40 dB below the level of the carrier, or a maximum of 50 mW.

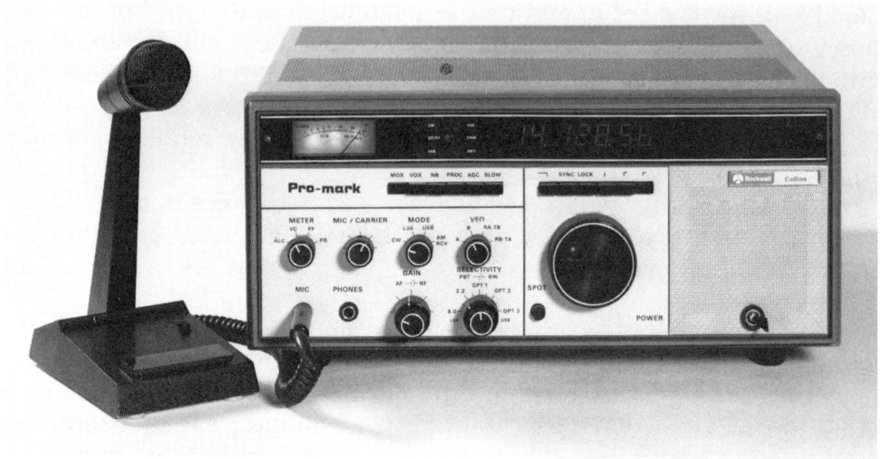

The Collins KWM-380 is a synthesized transceiver with all amateur band transceive capability and general coverage receive capability. (Courtesy of Collins Group of Rockwell International.)

Many modern solid-state transceivers are capable of providing this level of carrier and unwanted sideband suppression. Many older tube-type transmitters and transceivers were capable of providing this level of suppression only if they have been recently adjusted. The high-quality crystal filters and the double balanced diode mixers used in many new transceivers will provide the high level of suppression for the life of the equipment.

Many transceivers have an RTTY position in the mode switch which can provide some important changes in the operation of the transceiver. First, the BFO is retuned so that the beat frequency corresponds to the standard mark and space tones. Without changing the BFO frequency, it would not be possible to use the narrow-bandwidth filters for RTTY. Normal lower-sideband operation places the voice frequency within the passband of the IF filter. CW filters are usually designed to use the same BFO frequency but to place the signal about 800 to 1000 Hz from the BFO frequency. Since the desired beat frequency for radioteletype is 2125 to 2195 Hz, the BFO should be moved up about 1200 Hz.

As always, the transmitter, or transceiver, should be capable of operating for long periods of "key down," and this capability should be so stated.

RTTY operation is a good point at which to discuss the pros and cons of synthesized transceivers. There are many transceivers that are advertised to be synthesized when they merely use a phase-locked local oscillator. A synthesized transceiver is controlled from a digital signal and is tuned in discrete steps. Phase-locked local oscillators are slaved to a conventional VFO that has all the characteristics of that type of device. The frequency drift of a synthesizer is essentially zero, which is an important characteristic for certain types of operation, especially RTTY. For good printing, the RTTY signal should be tuned to within 15 Hz of the correct value and remain within 15 Hz of that value. This requires stable equipment and usually an occasional retuning. This extreme stability would make the synthesized transceiver seem the obvious choice, but in practice it can be very difficult to use.

Every synthesizer has discrete frequencies, which in the case of amateur equipment is often 100 Hz. It can be seen that it could be impossible to tune an RTTY signal to within 15 Hz with a synthesizer that has 100-Hz channels. Most synthesizers have a "vernier" control that can span the distance between channels, but the two-knob tuning scheme could be more annoying than the occasional retuning required in a conventional transceiver. Some transceivers have 10-Hz steps, which would give sufficient resolution for normal RTTY operation. Designing a synthesizer that will provide 10-Hz steps plus the necessary electronics to program the synthesizer without producing serious spurious signals is a very difficult job. Synthesizers with 10-Hz resolution are, of necessity, very expensive and are usually found only in high-end equipment.

Traffic operators would require full break-in operation. Some transceivers are advertised to have "semi-break-in," which is often nothing more than a VOX-like system for CW, which is usually too slow to provide any break-in advantage. In fact, in most semi-break-in transceivers, the systems in use are the nomal VOX circuits. It is difficult to specify the performance of a break-in system, and few transceivers attempt to do so. If break-in operation is a requirement, it would be advisable to actually operate the transceiver on the air, to become familiar with the break-in system.

For the DX hound, a separate VFO would be a must for split-frequency operation. Also, for DXers, who often work near the band edges, a digital readout should supply the additional dial accuracy required to stay within the band. For those avid DXers who may make trips to foreign countries on "DX-peditions," a power requirement of 220 V and 50 Hz may be necessary.

The serious contest operator requires the usual features that any serious SSB or CW operator requires plus the extra cooling and high-duty-cycle requirements that the RTTY operator needs. Operating a long contest can put a strain on a transceiver if it is not designed for heavy use. Many transceivers have accessory cooling fans that can be added to allow almost continuous operation without a dangerous heat buildup.

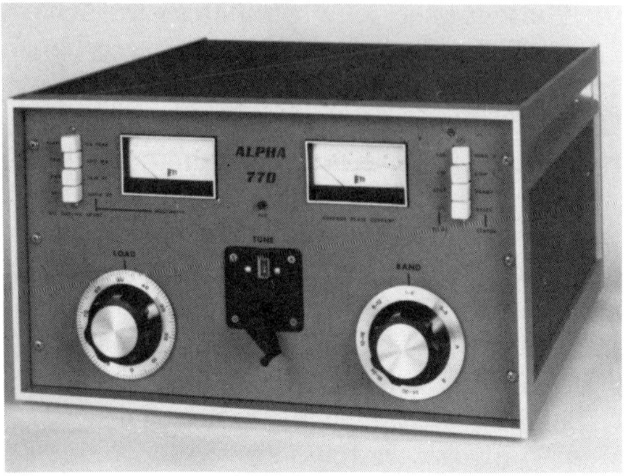

The Alpha 77D is a full legal power amplifier capable of operating all of the difficult modes without derating. (Courtesy of Ehrhorn Technological Operations, Inc.)

There are no magic numbers that can be used as minimums for selecting a receiver or transceiver. It cannot be stated that 0.2-μV sensitivity for CW is required or that a 300-Hz selectivity is a requirement. How a receiver or transceiver will function under certain adverse conditions is a function of many of the specifications. If a list of specification minimums and maximums were established and two receivers were selected that satisfied the minimums and maximums, it is possible that the two units could have two different performances under certain conditions. Most amateur receivers and transceivers have very impressive published specifications. It is the specifications that are not published that tell the real story. When selecting a receiver, it is better to consider the opinions of owners than simply to compare the published specifications.

5.4 USED EQUIPMENT

To many amateurs on a limited budget, a new transceiver is beyond reach. There are many pieces of used equipment available for a fraction of what they originally sold for or what they would cost to replace. However, when buying used equipment, extra caution, beyond selecting the right kind of equipment, must be taken. Several rules should be followed when investigating used gear. Unless they are unrealistically inflated, used equipment that was expensive when new is a good buy as used equipment. No cost was spared in providing a high-quality product with good electrical design that was constructed to last a long time. In some cases, this philosophy backfired, leaving the sorry buyer with a high-quality, long-lived piece of obsolete equipment. Most notable historically were high-power and high-quality VHF equipment for AM introduced just before the advent of VHF repeaters. This helped push some manufacturers into bankruptcy, and some excellent small companies that produced high-quality equipment failed because of the small market for the quality equipment. However, some smaller companies failed because of inferior products. Therefore, before buying any equipment manufactured by a now-defunct company, determine why that company failed. It would not be a bad idea to investigate the manufacturer of any used gear you are thinking of purchasing.

When buying a used piece, be sure that an instruction book, including a schematic, goes with the price. The cost of a replacement book, if one can be located, is always very high. Never buy a used piece of radio equipment without first looking at it. Do more than spin the knobs; open the cabinet and take a good look inside. If it appears that there are modifications, be sure to ask the owner about them. Many modifications, if carefully done and documented, are well worth the trouble and enhance the operation of the equipment. This is especially true of modifications that are suggested by the equipment manufacturer. Many transmitters or receivers were updated to later production models by installing factory-supplied modification kits in accordance with the manufacturer's instructions. If this is the case, the instructions should have been incorporated into the instruction book.

The equipment should be operated and some tests made. It is difficult to make all significant tests when buying a used transmitter or transceiver, but at least a simple functional test using the internal meters should be performed. Other tests using an external wattmeter or signal generator will be even more valuable. It is not fair to perform a complete test on equipment that is relatively old and expect it to meet all factory specifications. On the other hand, if the selling price is high because the seller claims that the equipment is in perfect "like-new" condition, he or she should not be upset if you suggest that a complete check of the equipment be performed. The complete test will probably show that the "like new" refers to the outward appearance and not necessarily its electrical operation.

Buying used equipment from a dealer that takes used equipment in on trade does not necessarily ensure any more than buying directly from the previous owner. In fact, buying directly from the previous owner allows you to determine how the owner kept the equipment maintained, how he or she liked using it, why it is being sold, and so on. Used equipment on a dealer's shelf has been cleaned so that all traces of a sloppy or careless owner have been erased. At many dealers, there are no guarantees on used equipment, so buyers are on their own anyway. Which brings up the topic of guarantees and warranties.

A guarantee is something that should be read very carefully. Be sure that you understand just what is covered and what has to be paid for by the customer. Electronic equipment is notorious for having a lot of inexpensive parts that cost a small fortune to replace. If a warranty makes the buyer responsible for the cost of labor while the manufacturer supplies the parts, this is no warranty. Most warranties specifically state that both parts and labor are covered. In some cases, certain parts are excluded from the warranty, such as vacuum tubes and power-output transistors. This is unfortunate, because these are parts that should not fail within the usual 90-day warranty period and are expensive if the cost has to be borne by the owner. The only protection the new owner has in these areas is the reputation of the manufacturer.

5.5 KITS

A considerable sum of money can be saved if amateur radio equipment is built from kits offered by several manufacturers. Although there is a saving of money, there is a considerable investment required in the form of time. Most manufacturers of kits will provide some form of estimate of the time required to completely build and align one of their products. If the instructions are followed carefully, the resultant equipment is usually of high quality and will provide many years of operation.

One major kit manufacturer has a policy of selling the instruction and assembly manual to prospective buyers for a nominal price and refunding the price upon purchase of the kit. Having the assembly manual for a specific kit can give some important insight into the amount and kind of work required to complete the kit.

5.6 SELECTING VHF EQUIPMENT AND FM TRANSCEIVERS

The process for selecting VHF equipment is carried out in a similar fashion as for high-frequency equipment. The same process of determining price ranges and evaluating manufacturers by asking and listening can provide similar valuable information about amateur VHF equipment. The most common piece of VHF communications equipment bought today is the VHF FM transceiver. Most VHF equipment sold is synthesized and has available several hundred channels with two or more repeater offsets. Some crystal-controlled transceivers are still available and would be purchased for a specialized purpose or when the budget is very tight and the absolute minimum is to be spent on equipment. Assuming that every channel in a crystal-controlled transceiver requires two crystals, one for transmit and one for receive, about 10 channels would cost as much as the average cost differential between a synthesized transceiver and a crystal-controlled unit. Some later designs require only one crystal for each channel and would take about 20 channels to make up the difference in cost. Although most operating will take place on only a few channels for the average amateur, there are times when the flexibility of a synthesized transceiver will be invaluable. This is especially true when traveling.

The power level of the VHF transceiver is not usually important unless the unit it to be used from batteries that are not being charged, such as would be the case if the transceiver is used as a portable unit with a battery pack or on a boat, where there is no charging system in operation. In this case, it is desirable not to have excessive power or a switch that can select the power output. Ten watts is usually sufficient for adequate repeater and local simplex operation. Greater power output is desirable only if there are known trouble spots within the area of use or if the repeater is at a great distance. DXing is not a welcome activity on the repeater portion of the band, as it tends to interfere with co-channel repeaters. Simplex-channel DXing is permissible but not usually fruitful, because of the high level of activity on those channels. Most long-distance work on the VHF bands is carried on SSB and CW, so high power is not usually useful for FM transceivers.

Because of interference from co-channel repeaters, some systems use either a tone burst or inaudible squelch system, to prevent interference from the other repeater. These tone-activated squelch systems are offered as an option or even as standard features in some FM transceivers and are a worthwhile option.

Most of the specifications used for the characterization of a high-frequency receiver or transceiver are used in the description of VHF FM equipment. There are some significant differences. First, FM systems differ in behavior from AM or single-sideband systems. Sensitivity which was specified for a particular signal-plus-noise to noise ratio for SSB is specified as the minimum input signal that will provide a certain amount of quieting, which is the reduction in background noise by a received carrier. Unlike the SSB or AM sensitivity measurement, no modulation need be applied to the signal used to measure this parameter. Specifications tied directly to modulation are audio distortion and audio power output. An effect unique to FM systems is known as the *capture effect.* This allows two signals to be present on the same channel; the receiver demodulates one without distortion while the other signal is unheard. The stronger signal is always the audible signal, while the lower-level signal causes no objectionable interference. The minimum difference in decibels to produce this effect is known as the *capture ratio.* The lower the number, the better the capture ratio. Usually, on a repeater channel, this effect will not be important; however, on a simplex channel, where the interference is heavy, such a characteristic would be very valuable. A good capture ratio is usually an indication of good gain distribution and good limiters, which also help the suppression of ignition-type noise and produces good sensitivity, both characteristics important to FM equipment, especially that used for mobile operation.

Many synthesized FM transceivers have digital computers or other circuits to allow the storing and recalling of certain frequencies, as well as a scanning function for automatically tuning the transceiver. Most of these functions do not cost the manufacturer much in terms of additional hardware. Usually, the addition of such functions involves providing the necessary switches and pushbuttons to invoke these features and additional programs in the microcomputer. Therefore, the cost premium should be minimal. A transceiver that is considerably more costly and appears to have only these features is not a good buy. Some amateur-related operations, such as MARS and CAP, involve transmitting outside the amateur band. In some transceivers, the trans-

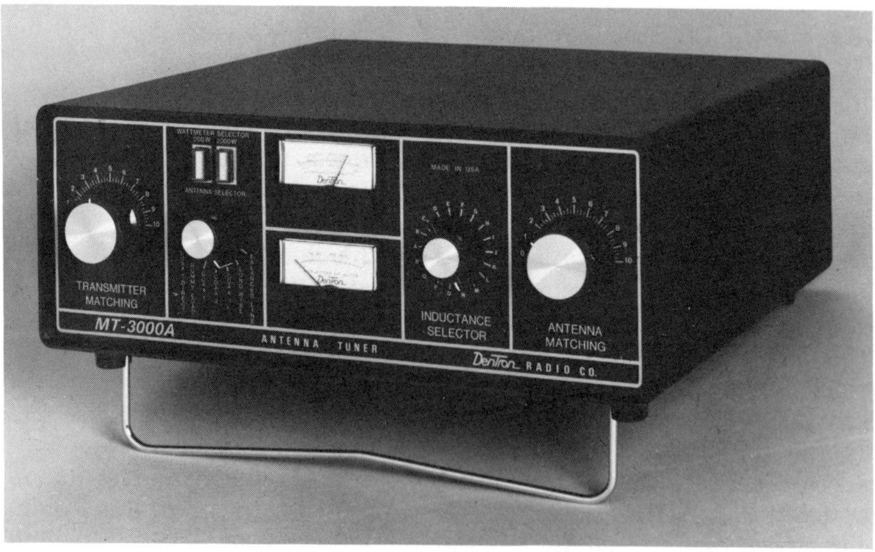

The Dentron MT-3000A antenna tuner can match just about any type of antenna at the maximum legal power. (Courtesy of Dentron Radio Co. Inc.)

mitter is disabled so that out-of-band transmissions cannot take place. If out-of-band transmission is to be used, it would be a good idea to check with the manufacturer and other users of the equipment to see if it is possible to use the equipment out-of-band. In some cases it will involve disabling the transmit inhibit, and in other cases, the synthesizer is not capable of operating out of the band at all.

UHF FM equipment is very similar in appearance and operation to 2-m FM transceivers. Very few synthesized UHF transceivers are currently available for the 450- and 220-MHz band. There are no available transceivers for any bands above 450 MHz, although the 1215-MHz band is potentially useful for mobile operation.

An important fact to keep in mind whenever purchasing UHF equipment is that the repeater offset frequencies on the UHF bands tend to vary by quite a bit, which almost requires separate frequency selector switches for both transmitter and receiver or a large number of repeater offset selections. Most crystal-controlled transceivers require a separate crystal for both transmit and receive which would allow any repeater offset to be programmed by simply ordering the proper crystal.

There is available for both UHF and VHF so-called multimode transceivers that offer SSB, AM, CW, and FM operation. They are quite expensive and considerably larger than an FM-only transceiver. The majority of the activity on the VHF and UHF bands is FM, while the majority of serious DX work and contest operation is CW and SSB. The decision must be made on the basis of DX and contest work if the expense and extra size and weight is worth the additional modes.

5.7 SELECTING A HIGH-POWER LINEAR AMPLIFIER

Most amateurs eventually add a linear amplifier to their station. There are many good reasons for increasing the station power to nearly the legal limit, and there are good reasons for not doing so. First, it is possible to operate quite successfully and make contacts throughout the entire world with much less than the legal limit. This is assuming that conditions are good and that the station uses a good antenna. Many amateurs use high power to combat interference. This is self-defeating, for if the interfering station were not using high power, the interfered-with station would not have to resort to high power. However, when conditions are bad or there is a lot of atmospheric noise, such as 80 m in the summertime, high power can make the difference between making a contact or not.

Selecting a high-power amplifier should be done with as much care as selecting a transmitter or transceiver. There are not many performance features offered in an amplifier, so most of the selection will be based on price and quality.

The first and foremost decision to be made is what power-output level the amplifier will have. Most amplifiers are capable of producing the maximum legal limit on CW, and many can produce the legal limit on SSB. For CW, 1000-W dc input power is the legal limit, while for SSB, 2000 W PEP is the limit. Many amplifiers are capable of supplying 1200-W PEP input power for SSB and 1000-W for CW. Using a linear amplifier for any power lower than this is not worth the cost of the amplifier. Most transceivers supply an output power on the order of 100 W. The output power from a 1200-W PEP input amplifier is about 600 W, which is only 7 dB above the driver power. Of course, the maximum legal limit is only 3 dB above that level. If the power of an amateur station is increased by the use of a linear amplifier, and price is not a very critical consideration, it might as well be increased to the limit.

It was suggested earlier that equipment from only one manufacturer be purchased, to ease the job of interconnecting the individual systems. This is also true of linear amplifiers, especially if a full break in operation is used. Most amplifier interfaces are nothing more than connecting the RF out from the transceiver to the amplifier, connecting the linear amplifier to the antenna, and connecting two wires for a T/R switch. There are other interconnections that are desirable and are not always available if the equipment is that of two different manufacturers. One is the use of ALC. This allows control of the output power by sensing the power output and feeding back, to the exciter, a control voltage to adjust the gain in the exciter. This method produces the most effective ALC action without generating excessive distortion. Another feature that may be lost if manufacturers are intermixed is the use of break-in operation. True break-in usually requires a sequenced operation of the system relays. Unless the relay circuits are tailored specifically for the amplifier involved, the T/R relay in the amplifier may open or close at the wrong time.

Luckily, excluding the loss of break-in and ALC, most amplifiers will interface with practically any transceiver or transmitter. In some cases, ALC and break-in connections are provided on some amplifiers for popular transceivers or transmitters.

There are basically two categories of linear amplifiers as far as power output is concerned: the 1200-W PEP and the 2000-W PEP class. When selecting an amplifier to be used primarily for SSB operation, these are the power inputs that will be supplied. When using a linear amplifier for RTTY or slow-scan television, or even for CW, the power rating of the amplifier can vary considerably. When selecting an amplifier for one of these modes, such as RTTY, make sure that the RTTY power is specified. It cannot be assumed that the RTTY power is the same as the CW power input.

Another important power rating to investigate when selecting an amplifier is the drive-power requirement. Obviously, the drive requirement to drive the amplifier to its full power output cannot exceed the power supplied by the transceiver or transmitter. Do not forget that the transceiver power output, not the input, must be sufficient to drive the amplifier. Most grounded grid amplifiers require about 100 W of drive to supply full power output. Since most transceivers are about this power rating, the driver requirement is seldom any problem. It is a good idea to match the drive requirements closely to the transceiver. A transceiver that cannot supply the required power will not drive the amplifier to full power output. A transceiver that has excessive drive must be either attenuated or the level control on the transceiver must be decreased. Both of these techniques are undesirable. Attenuating the power is a waste of energy, and lowering the level control causes the exciter to be operating below its optimum level.

Low-power exciters can offer some special problems. There are some low-power transmitters with less than 100-W output power. These units will not drive most of the linear amplifiers available on the market. There have been linears that were grounded-cathode-grid-driven amplifiers having a drive requirement of less than 10 W. These amplifiers are not as efficient as those described in Chapter 2, but are the only type of amplifier suitable for low-power transmitters or transceivers.

All linear amplifiers do practically the same thing; that is, they provide about 10 dB of gain to a level of 1000 W. How is it, then, that there can be such a wide variation in price? The reasons are varied. The ability to operate slow scan or RTTY will dictate heavy-duty components and is the reason for the increased cost of units specified for that type of operation. There are other differences in cost for amplifiers with similar specifications. The quality of the components in the amplifier is the overriding factor for high-priced amplifiers. For example, a high-priced amplifier would probably have vacuum-variable capacitors, whereas the lower-priced amplifier will have air capacitors. The voltage breakdown for vacuum capacitors is usually much higher than that for a comparable air capacitor. Although both amplifiers can operate without breakdown at the legal limit, the vacuum variable will be less likely to arc over if the amplifier is severely mistuned. This overrating of components would be evident in other parts of the amplifier. This means that the more expensive amplifier would be less likely to be damaged if it is mistreated. Is the extra cost worth it for this characteristic? It is up to the buyer to decide if he or she is willing to pay the price for the higher-quality amplifier.

5.8 SELECTING OTHER STATION ACCESSORIES

Aside from the transmitter, receiver, and high-power amplifier, there are a large number of accessories that are useful for the amateur station.

One very valuable accessory is an antenna tuner. There have been endless arguments among amateurs over whether or not an antenna tuner is necessary. One thing can be said about the tuner; having one cannot hurt. Vacuum-tube transmitters with pi-network output stages can load into antennas that were not a reliable 50 Ω. For these transmitters, the antenna tuner may not be necessary. Modern solid-state transmitters have no output tuning and will operate properly only into a 50-Ω load. In fact, many solid-state amplifiers have built-in SWR sensors and cause the power to be cut back whenever the SWR exceeds a preset value. The use of an antenna tuner with a solid-state transmitter or transceiver is highly recommended.

Just what can an antenna tuner do? First, it can transform impedance, and second, it can tune out reactance. That is, an antenna of any impedance, provided that it is not outlandish, can be transformed to 50 Ω resistive. In addition, since the tuner is a tuned circuit, some attenuation to harmonics is provided.

When selecting a tuner, several important things must be considered. First, the power capability of the tuner must be investigated. A tuner capable of handling 1 kW would be the safest purchase. Second, how versatile is the tuner? Many tuners can handle balanced feed lines, random wires, and coax. A very versatile tuner can match practically anything to 50 Ω.

There are many other accessories for the amateur station, too many to discuss. When buying a station accessory, use the techniques outlined for selecting a transmitter or transceiver.

6 MODERNIZING AND RESTORING OLDER AMATEUR EQUIPMENT

6.1 PURCHASING AN OLDER UNIT

There are many fine examples of older equipment still on the air. There is a brisk market for used and older equipment from dealers and private sales where a transmitter or receiver may be purchased for a fraction of the original cost. The techniques for purchasing a used receiver, transmitter, or transceiver as outlined in Chapter 5 apply to equipment less than 10 years old. Although there have been many technological advances in recent years, amateurs on a limited budget can usually obtain a station of superior performance if they are willing to invest time to modernize an older piece of equipment.

Any amateur transmitter or receiver, regardless of how old, can be used successfully on the air without any modifications or modernizing at all. Certainly, 20, 30, or even 40 years ago amateurs were communicating successfully with equipment available at that time. Of course, the modes of operation used many years ago were somewhat different than those used today. Most older equipment was capable of operation on CW and AM phone only. Although AM is still legal, it is being used very little, if at all. Of course, CW continues to be a popular mode on all bands. With a little restoration, an unmodified antique amateur station can provide reliable and enjoyable CW operation, even in today's crowded bands. Very old equipment will not provide the measure of performance that a new transceiver, transmitter, or receiver will provide. In many cases, it would be desirable to add extra circuits or modify the circuits of a receiver or transmitter to improve the performance. The extent to which modifications can be made and the performance gains the modifications obtain will depend on the experience level of the person making the changes. Modifications as simple as replacing the rectifier tube with silicon diodes can be handled by the novice operator and can provide some measure of performance improvement.

More complicated modifications, if not done properly, can actually degrade performance. It is possible to modify a receiver or transmitter to such a degree that the entire vacuum-tube circuitry is removed from the unit using only the original cabinet, coils, variable capacitors, and the dial mechanism. This technique has been used in more than one instance and, if done by a very experienced amateur, produces a receiver or transceiver with performance rivaled only by the newest or higher-priced units. Most amateurs investing in older equipment are less experienced amateurs looking for a way to save money on the equipment investment. Therefore, most modifications will be small and easy to do.

The discussions of Chapter 5 dealing with the selection of an amateur transmitter receiver or transceiver apply to the purchase of an older unit. However, there are some additional precautions for selecting a unit that will be repaired or modified. The biggest watchword is price; do not overpay for old equipment. The purpose of buying older equipment is to save money. In some instances, the seller will claim that the equipment is a "collector's item." There is a very small market for collectable amateur equipment, and pieces of equipment falling into that category are very old, usually more than 40 years old. Equipment that old is not desirable, for several reasons. First, the price is liable to be inflated because of its collectable value. Second, parts, especially vacuum tubes, will be very difficult to locate. Finally, the older a piece of equipment is, the more crude the technology was and the unit will rarely perform well.

When purchasing any piece of used equipment for modification, it is absolutely necessary that the instruction book be included with the price. Under no circumstances should you obtain a piece of equipment if the schematic is not available or included with the price. Instruction books for old transmitters and receivers are available, but the prices tend to be very high, if they can be found at all.

Also check to be certain that the unit is complete. If any parts are missing, it may be difficult, if not impossible, to locate a replacement part, particularly for parts that are unique to the unit. If the unit is incomplete, it would be advisable not to purchase, regardless of price.

When purchasing an older unit, it is wise to know what kind of equipment was generally available at the time of manufacture and determine if the receiver or transmitter was a high-performance or low-priced unit. It is generally advisable to purchase only older equipment that was high-priced. There is no sense in purchasing an old, cheap unit, and purchases should be limited to the more expensive units. An old *Radio Amateur's Handbook* or old magazines can provide an overview of the type of equipment sold during the time of manufacture. The specifications can be compared to the other units offered for sale at that time.

One type of equipment that should always be avoided is the transformerless "ac/dc" receiver. A few receivers were manufactured with this type of power supply. Most receivers using this technique were not suitable for amateur use and were purchased by those interested in shortwave broadcast. There were a few models that had amenities for the amateur and appear occasionally on the used market. The problem with this type of equipment is that one side of the ac line is internally connected to the receiver circuits, making it nearly impossible to add safely external equipment such as preselectors or Q multipliers.

6.2 REPAIRING AND ALIGNING EQUIPMENT

After the transmitter or receiver has been purchased, the first step in making the equipment useful again is to assess exactly what does or does not work. The worst thing one can do is to plug in the unit, turn on the switch, and duck. The only smart move in this procedure is ducking. The very first thing that should be done is to remove the cabinet bottom plates, and other covers so as to get a good look at the inside. Look for obviously destroyed or burned parts, which should be identified and replaced. If possible, what caused the part to fail should be determined and the cause repaired. It is also desirable to look for modifications and to trace out the modified circuits to determine what each addition or change does. Finally, the equipment should be turned on. A very handy device to have when restoring older equipment is a variable transformer (Variac) with an ammeter. With the variable transformer, the line voltage can be raised gradually while the input current is monitored carefully and the chassis inspected for signs of burning or arcing.

Once the full line voltage has been applied, it can be determined just how much of the receiver or transmitter is functioning and where repairs must be made. If one is lucky, a receiver will immediately begin to deliver audio to the speaker and perhaps even receive stations. The broadcast band, if it is included, is a good band to start the receiver testing, as signals are always present and strong. When testing a transmitter, the lowest band, usually 80 m, should be used and a dummy load should be connected to the output jack.

Unless the equipment has been severely damaged, it is usually possible to duplicate factory specifications provided that those specifications were honest in the first place. If modifications are desired, it is best to start by placing the receiver or transmitter in a complete operating condition.

Repairing the older receiver or transmitter may be a problem for younger amateurs who cut their teeth on transistors and integrated circuits, while old timers will plunge in directly, feeling right at home. Those who require a refresher course on vacuum-tube technology should consult older *Radio Amateur's Handbooks* for discussions of vacuum-tube circuits. Fortunately, amateur radio equipment used pretty straightforward circuits that are discussed thoroughly in the handbooks.

After the unit is made completely functional, it should be brought up to factory specifications. This may require nothing more than a realignment, but in most cases, a certain number of repairs will be required. If the receiver lacks sensitivity, check the vacuum tubes in the RF and IF amplifiers. If the sensitivity does not respond to new tubes, check the operating voltages and parts values. It is not uncommon for parts to change value over a period of time.

All receivers should receive a complete alignment. The proper equipment and technique should always be used to align the receiver; a receiver should never be aligned "by ear." This will usually end up causing more damage than good. Some instruction books included an alignment procedure, which should be used. Lacking such a procedure, a general procedure may be determined from a textbook on radio servicing or by adapting a procedure from another model.

CW/AM transmitters are aligned without the aid of a signal generator or any specialized equipment. Single-sideband equipment, on the other hand, should be aligned carefully using only the proper equipment. Failure to properly align a SSB transmitter can cause spurious emissions.

Some components used in earlier communications equipment are prone to failure and should be checked before placing the equipment into regular service. Electrolytic capacitors are the most likely components to fail because of aging. If a receiver has an audible ac hum, the filter electrolytic capacitors should be checked. This may be accomplished by placing a good electrolytic capacitor across the suspected capacitor and listening to the ac hum level. Any significant drop in the hum level indicates a faulty capacitor. Be sure to discharge the test capacitor after use, as it can retain a dangerous amount of charge.

Tubular capacitors tend to become leaky, which can cause a number of problems—most often, distortion and overheating in audio amplifiers. Any component that has been visibly damaged or shows discoloration from overheating should be replaced with a new unit. Do not try to replace any component with an exact type of part, as many of the parts found in older equipment are not made in exactly the form they were in the past. Newer parts are superior to the older type of component. For example, paper capacitors that were used in large numbers are no longer made. All paper-type capacitors may be replaced with plastic film and in most cases, with ceramic capacitors of the same ratings.

6.3 ADDING MODIFICATIONS

After the transmitter or receiver is fully operational and nearly up to original specifications, an assessment should be made of its strengths and weaknesses. Many receivers were noted for their sensitivity, while others were noted for their selectivity, dial accuracy, stability, and so on. Compare the original specifications with those of a modern receiver. Many specifications will be close to those advertised today and others will be sorely lacking. It would be best to modify the transmitter or receiver to strengthen its weakness rather than improve on a strong point that cannot be used fully because of an inherent weakness. For example, if a receiver had good selectivity but suffered from excessive frequency drift, it would be foolish to further improve the selectivity, since the additional selectivity would serve to make the frequency drift all the more noticeable.

Many receivers were modified by their owners according to instructions supplied by the manufacturer. In some cases, these modifications changed the receiver circuit to be identical to later production units. In other cases, the modifications improved the receiver performance but did not modify the circuit to any production configurations. In either case, the modifications should be documented by factory-supplied diagrams and in some cases by factory-supplied parts.

Some weaknesses were common to a number of receivers. One of these was frequency drift due to the wide temperature variations due to the heat buildup inside the cabinet. There are numerous sources of heat in a receiver, but two of the larger sources can be easily removed.

The easiest source of heat to remove from the receiver is the power rectifier tube. This tube may be replaced with silicon diodes with very little modification. Figure 6-1 shows a portion of a communication receiver schematic and Figure 6-2 shows the rectifier tube removed and

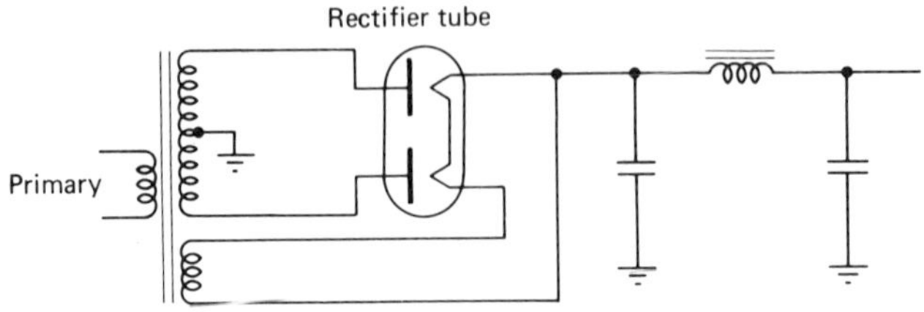

FIGURE 6-1. Typical power supply rectifier from an old receiver.

replaced with silicon diodes. Any 1/2-A diode having a peak inverse rating of more than 1.4 times the total secondary voltage will be satisfactory. Notice that the 5-V rectifier heater windings are no longer required. Do not cut these wires short, as they may be used at a later time to generate power for other solid-state accessories.

The forward voltage drop of the silicon diodes is considerably less than the vacuum rectifier tube, and the resultant supply voltage may be larger than before. Most receivers can handle the increased voltage without any modification, since in most receivers the supply voltage will be greater when the receiver is first turned on. It is a simple matter to measure the voltage at the filter capacitors and compare the measured voltage to the rated working voltage to be certain that the parts are not being overworked. If the voltage is excessive, a series resistor can be inserted between the diodes and the first filter capacitor to lower the voltage. A 100-Ω 1-W resistor would be a good starting point. It is necessary only to lower the voltage to below the working voltage of the capacitors plus a few more volts for good measure.

A second source of heat in the vacuum-tube receiver is the audio power amplifier. The power amplifier can either be removed from the receiver or replaced with a solid-state amplifier. The audio may be removed from the receiver by connecting a lead from the center of the audio gain control and routing the signal to an external amplifier. A small stereo amplifier of a few watts would be an ideal amplifier for this purpose. The stereo amplifier could be used as a station audio control

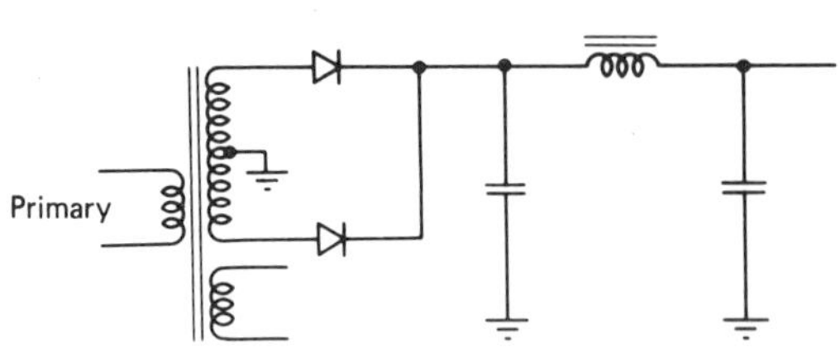

FIGURE 6-2. Silicon diodes used to replace the rectifier tube of an old power supply. The 5-V winding is no longer needed.

system by mixing several sources, such as other receivers, and amplifying them through a common speaker system. The base and treble tone controls may be used as a rudimentary filter. A more elegant method of removing the audio amplifier heat is to replace the vacuum-tube audio amplifier with a solid-state unit. Several integrated-circuit audio amplifiers are suitable for a communications receiver, including the LM-380 schematic as shown in Figure 1-20. A power source will have to be created to power the amplifier. Twelve volts, at about 1/4 A, is required for the LM380 circuit. A small 12-V filament transformer may be mounted on the receiver chassis and the primary wired in parallel with the receiver power transformer, so that the 12-V supply will be switched with the receiver power switch. A full-wave bridge rectifier and a 1000-microfarad (μF) capacitor will provide sufficient current for the amplifier.

Most older receivers used a simple crystal filter for CW and relied on the IF transformers for AM selectivity. Receivers of modern design use either crystal lattice filters or mechanical filters for SSB selectivity. The bandpass characteristics of these filters are far superior to those of the IF transformer. If the older receiver has a common IF, such as 455 kHz, great improvements in selectivity can be obtained by installing a modern filter. Figure 6-3 shows a mechanical filter installed in a vacuum-tube receiver. It is important that the mechanical filter be terminated with the proper capacitance in order for it to have a good passband ripple.

Unless the receiver was low in gain before the mechanical filter was installed, there should be sufficient gain to make up for the loss in the mechanical filter. Because mechanical filters are generally expensive, ceramic filters may be preferred for improved selectivity. Some of the larger ladder-type filters could improve the selectivity of a receiver for considerably less expense; however, the much lower impedance level of the ladder filter makes it somewhat more involved to interface with the

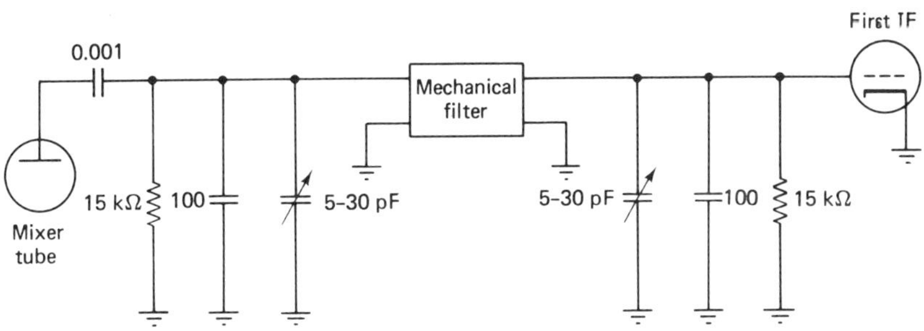

FIGURE 6-3. Mechanical filter inserted between a mixer and the first IF amplifier in a vacuum-tube-type receiver.

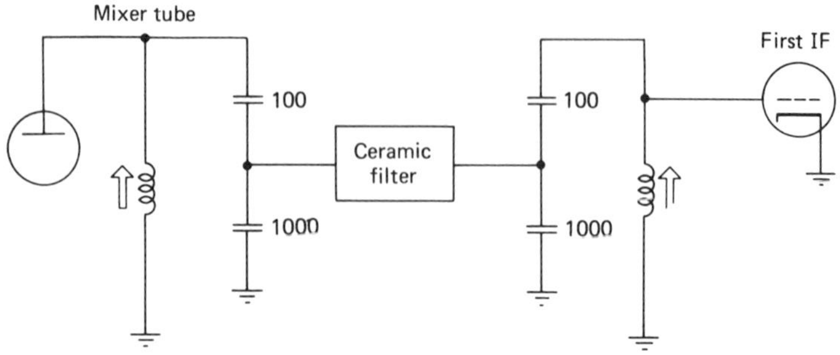

FIGURE 6-4. Ceramic filter installed in a vacuum-tube receiver. The low impedance of the filter requires a slightly different matching network than the mechanical filter.

vacuum-tube circuits. Figure 6–4 shows a vacuum-tube IF amplifier modified to accept a ceramic filter.

Improving the selectivity of receivers using an IF other than 455 kHz is difficult. It may be helpful to determine if another receiver manufactured by the same or a different manufacturer used the same IF. If this is the case, an IF filter from a later receiver, if it is available, may be installed in the older receiver.

The addition of a product detector can improve the CW and SSB performance of a communications receiver. Like most modifications of older equipment, it is best to find a description of the modification applied to the specific receiver in question. Lacking that, the product detector shown in Figure 1–18 may be added. Since the impedances of the solid-sate product detector are much lower than the equivalent vacuum-tube circuit, some impedance matching must be done. This is best left to the experienced amateur.

Unlike the diode detector used for SSB or CW detection, the product detector can have considerable gain. This implies that the detected audio could be excessive for the audio amplifier. A switch must be provided to switch between the product detector and the original diode detector.

Numerous journal articles have described internal modifications for popular transmitters, receivers, and transceivers. Most of the modifications were vacuum-tube additions for vacuum-tube receivers and solid-state additions for transistorized equipment. A search of all the major amateur radio magazines or a letter to the manufacturer can result in several easy-to-install and worthwhile modifications. It is helpful to know the date of introduction and search the year-end indexes for suitable articles.

A number of ills of the older receiver can be corrected with external add-on circuits. Inadequate sensitivity and image rejection are two ills that may be improved by the use of a preselector. Single-conversion receivers using a 455-kHz IF had notoriously poor image rejection on the higher-frequency bands, especially 10 and 15 m. In addition, the sensitivity of a receiver usually dropped off above the 14-MHz band, becoming worse at 10 m. A preselector combines a tuned circuit and an RF amplifier, which can improve the noise figure, sensitivity, and image rejection of a receiver. Several preselectors were manufactured in the 1940s, 1950s, and early 1960s. Although they improved the performance of a communications receiver, by modern standards these old preselectors themselves could be improved upon. Therefore, it is suggested that a home-constructed preselector using a good-quality field effect or bipolar transistor be used ahead of the older communications receiver. Such a preselector for the 10- and 15-m band is described in Chapter 9. Preselectors for the lower frequency bands are of little value, since very few receivers lacked sufficient sensitivity on those bands.

There are two external devices that can improve the selectivity of an inexpensive or older receiver. For CW operation, an audio filter can provide improved selectivity at modest cost. The audio filter with a 750-Hz center frequency, shown in Figure 6–5, should be inserted between the wiper of the audio gain control and the input to the audio amplifier. A switch is included to remove the filter from the circuit when it is not desired. The circuit could be installed inside the communications receiver, or audio leads may be brought outside the receiver and connected to the audio filter externally. This filter is not suitable for SSB operation or in conjunction with a crystal filter.

A Q-multiplier is an excellent device for improving the selectivity of an inexpensive receiver. They can be used as a peaking circuit or as a notch for removing undesired signals. There were several Q-multipliers manufactured and at least one is available in kit form. In addition, several circuits have been described in amateur magazines. More important than obtaining or building a Q-multiplier is knowing where to connect the device to the receiver. The usual connecting spot is the mixer plate, which can easily be found if the schematic of the receiver is available. This is one of many reasons for obtaining the schematic of any used equipment.

Crystal calibrators were available for most receivers as an extra cost accessory. A surprisingly few receivers supplied a crystal calibrator as standard equipment. The crystal calibrator is an absolute necessity for serious amateur work, with a multiple-frequency unit being the most useful. A crystal calibrator need not be built into the receiver, because interfacing the calibrator is a relatively simple task. It is possible to locate a crystal calibrator at a hamfest or flea market that will plug into the receiver in question and would probably be the easiest way to add a calibrator. However, the multiple-frequency calibrator is so superior to the older single-frequency 100-kHz calibrator that it is recommended

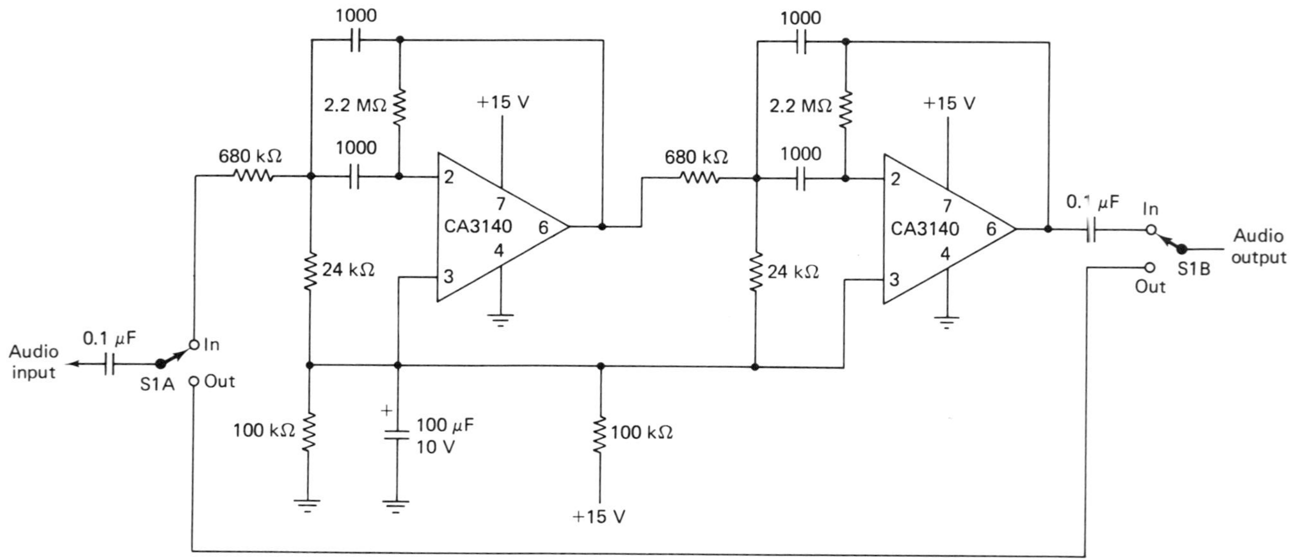

FIGURE 6–5. An active audio filter with a 750-Hz center frequency can be used to improve the CW selectivity of an older receiver.

in lieu of original equipment. A very simple and versatile calibrator is described in Chapter 9. No external power supply is required, and interfacing to the receiver is very simple. It provides three calibration points, for superior dial-calibrating ability.

It is generally more difficult to improve significantly the performance of an older transmitter. One very important point should be understood before purchasing an older transmitter. Most CW/AM transmitters cannot be converted to single sideband. There were some single-sideband adapters made for converting specific CW/AM transmitters, and these are the exceptions. However, the quality of the single-sideband signal is rather poor. Old CW/AM transmitters are still very useful for CW operation, especially for radio teletype. AM operation is rarely used anymore and the AM mode is virtually useless. Because of the AM mode capability, older transmitters were made to withstand long periods of key-down or carrier generation. Modern single-sideband transmitters, especially tube types, are not capable of extended periods of key down. But many of the older transmitters are capable of operating their full-rated CW power for very long key-down periods.

Some of the internal modifications outlined for the older receiver may be applied to transmitters. The tube-type rectifiers may be replaced with silicon rectifiers, thus removing the heat-generating vacuum-tube rectifiers. One-ampere diodes should be used for transmitters of 200 W or less. The peak inverse voltage should be 1.4 times the entire secondary voltage. Probably this will be over 1000 V, which requires that diodes be wired in series with equalizing resistors and capacitors, as shown in Figure 6-6.

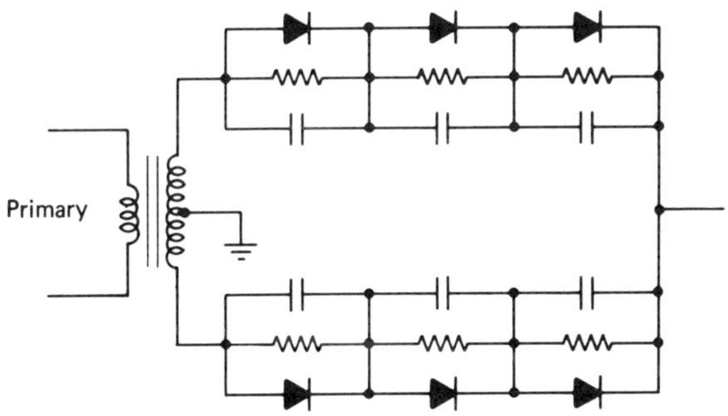

All resistors = 470 kΩ
All capacitors = 0.01 μF

FIGURE 6-6. When replacing rectifier tubes in a transmitter, high inverse voltages are required. Several diodes in series are usually required.

The voltage problems associated with this sort of modification in the receiver are also evident here. However, older transmitters were usually designed to withstand the voltage produced by the power supply when the transmitter was in standby. This voltage would be approximately the same as that produced by the power supply with silicon diodes installed.

Another heat-reducing scheme, similar to removing the audio amplifier in a receiver, is the removal of the modulator in an AM/CW transmitter. Very little AM operation occurs on amateur bands anymore, and the modulator in a transmitter is not required for CW operation. The simplest modification is to pull out the tubes in the modulator circuit. In some cases the modulator was a separate subchassis and can be removed completely. Even though the modulator is not used, power is being supplied to the heaters of all the modulator tubes.

The keying characteristics of many old transmitters can be improved by adding a time-sequenced grid block keying. The circuits required must be developed for each individual transmitter, depending on its tube lineup. Fortunately, many grid-block keying modifications have been described in amateur magazines over the years, and an applicable schematic can be found for practically any transmitter.

There are few external add-ons that can improve the performance of an older transmitter. Except for the normal additions that are used even on new equipment, such as linear power amplifiers, speech-processing equipment, and antenna-matching devices, very little is available that can breathe new life into an old transmitter.

One device that can be applied to an older transmitter is a single-sideband adapter. Several units were made during the 1950s and 1960s by major manufacturers. These units were phasing-type generators that generated the SSB signal at the operating frequency. Although these units suffered from many ills, they can be used and are still legal, even though the unwanted sideband and carrier suppression is poor by modern standards. If the adapter is used only on the 75- and 40-m bands, the unwanted sideband level and carrier suppression can be acceptable. The most important thing to do to a sideband adapter after purchase is to very carefully align the unit with the proper test equipment.

Most old CW/AM transmitters used the VFO/multiplier method and suffered from serious frequency-drift and tuning-rate problems on the higher-frequency bands. It is recommended that transmitters of this type be used only on the 80-, 40-, and perhaps, the 20-m bands. Most single-sideband transmitters, even many of the earliest ones, used the hetrodyne principle and do not suffer this problem. Perhaps the biggest drawback to an old single-sideband transmitter is inability to transceive. Most modern single-sideband transmitters are capable of tracking a companion receiver and thus operate much like a transceiver, with both transmitter and receiver being tuned by a single knob. Without such a feature, the transmitter frequency must be manually "spotted" by listening to the receiver. It is possible to build a device that would allow the receiver to be tuned from the transmitter, or vice versa. Some of these devices were written up in the literature and a small number were made commercially. It is probable that this feature will be difficult, at best, to install.

The most improvement to a single-sideband transmitter can be obtained by a complete and careful alignment. Many of the transmitter specifications, such as carrier rejection, unwanted sideband rejection, and distortion, are directly related to alignment. Follow the instructions in the shop manual and use the proper equipment. At least, in the case of a transmitter, there is no temptation to perform the tuneup "by ear."

6.4 RESTORING AND IMPROVING THE APPEARANCE OF OLDER EQUIPMENT

Before repairing or aligning a transmitter or receiver, the unit should be removed from its cabinet and brushed and vacuumed out. A small paintbrush can be used to dislodge dirt and dust from small crevasses, and a vacuum cleaner can be used to remove most of the dirt. Avoid using compressed air. The high-pressure air can damage capacitor plates and coils and cause extreme frequency errors. Switches and potentiometers may be cleaned with switch contact cleaner. The switches and potentiometers should be rotated while the spray is being applied so that the cleaner is worked into all the moving parts. Light household oil may be applied to the dial-drive shafts and gears. Be careful not to get any oil on the dial string or the dial itself. While the receiver is out of the cabinet, it is an excellent idea to replace the line cord with a grounded three-wire plug. Most old power cords have splits and cracks in the insulation and should be replaced.

The front panel of the receiver or transmitter may be restored using one of several techniques, depending on the damage and finish type. All panels should be washed using a spray type of cleaner followed with several water washings to remove all traces of the cleaner. Do not immerse the equipment in water; apply the cleaner and water with a small sponge, brush, or spray bottle. A small brush should be used on panels with wrinkle paint or other types of texture. Be particularly careful of the legends, logos, and other markings on the panel. It would be better to have these dirty than to wash them off.

Front panels having smooth paint and plastic parts may be further enhanced by using an automotive one-step wax and cleaner. Use standard household glass cleaner on the plastic and glass dial windows. The inside of the dial windows may be cleaned if they are accessible. Be careful while working around old plastic dial plates. They are often very brittle, and if they break, they cannot be replaced.

The knobs may be restored by removing them and scrubbing them with a small stiff brush, such as a discarded toothbrush. Dish detergent will remove most dirt from the knobs, and larger knobs may be further polished with the automotive cleaner.

Often, knobs used on older equipment have a very drab appearance even when they are clean. If this is the case, new knobs may be purchased. It would be advisable to try a knob or two before buying an entire set. A lot of old equipment looks downright silly with modern aluminum insert knobs.

The cabinets of older equipment usually took quite a beating and invariably, there will be dents and scratches. If the damage is slight, a good cleaning with detergent and a brush may be all that is needed. Moderate damage may be hidden with a new paint job. Smooth-finish cabinets should be sanded to remove the gloss and then painted with a coat of primer. All painting should be done with spray paint. A top finish coat and perhaps a coat of wax will give the smooth-painted cabinet a brand new look.

Restoring a wrinkle-finish cabinet is more difficult. If the damage is slight or moderate, the cabinet may be washed and sprayed with a primer without a finish coat. The flat appearance of primer tends to improve the appearance of a textured-finish cabinet without destroying the texture. Unfortunately, there is a limited choice of colors, usually black, gray, brown, and yellow. On the positive side, many cabinets were either gray or black.

If damage to the cabinet was extensive, body work will have to be performed. The first step is to completely strip the cabinet of paint, using a paint stripper or a sandblaster. The second step is to knock out the dents from the inside with a pin punch and a hammer. Be sure to back up the surface with a flat piece of iron or steel. After the dent is pushed out, any irregularities may be filled using automotive body putty and sanded smooth. These are exactly the same techniques and materials used by automotive body shops. The cabinet may now be primed and painted with either a smooth or a textured finish.

Install the transmitter or receiver in the finished cabinet using new hardware. The screws and washers used to hold the equipment in an older cabinet were usually nickel- or chrome-plated. If these have tarnished, it is nearly impossible to restore them to their original lustre, and new hardware can be just the necessary finishing touch to a complete restoration job. A good source for chrome- and nickel-plated hardware is a marine supply house.

6.5 VHF EQUIPMENT

Obtaining used VHF equipment presents a problem. Most old equipment is obsolete, and very old equipment, that is, equipment manufactured before 1950, does not exist at all. There were many models of VHF transceivers made during the 1950s and 1960s that can be obtained for very low prices. The sad explanation is that most of this equipment is of no value for use on the VHF spectrum. Most old transceivers were made for AM operation only. Unlike the HF spectrum, where AM died but CW lived on, the only mode that most VHF equipment could operate in was AM, and it has all but died on the VHF bands, leaving most transceivers useless.

There were other pieces of VHF gear that are still valuable and available at attractive prices. Any sort of receiving convertor may be used with a conventional receiver for SSB or CW reception. Transverters, which are converters for both transmit and receive, are available and are still very valuable for putting a high-frequency transceiver or transmitter/receiver on a VHF band.

Some older AM transceivers may be rescued by converting them to FM. An AM receiver can be used for receiving FM by using a tuning technique called *slope detection.* This method involves tuning the FM signal slightly off-center so that the signal is on the skirt of the IF filter. The FM modulation causes the signal to move up and down on the filter skirt and effectively changes the FM to AM, where it can be demodulated by the detector. Since the AM receiver has no limiters, the noise-reducing characterisitcs of FM will be lost. For a fixed station working local repeaters, this loss will hardly be noticeable, but for mobile operation, this loss will be a serious handicap. The AM receiver could be converted to FM by installing a discriminator and limiters. This would be an involved job for most transceivers, and would be hardly worth the effort.

Converting the transmitter to FM is not difficult if an external FM adapter is used. This device is nothing more than a crystal oscillator with an FM modulator. The microphone connects to the adapter and the internal AM modulator is disabled. The AM modulation removal must be done with care because in many cases, the AM modulator is also used for the receiver audio amplifier.

If a VHF transceiver is purchased for conversion to FM, several things should be kept in mind. First, transceivers with single-conversion receivers with a very high frequency IF should be avoided. There will hardly be sufficient selectivity for operation in the repeater portion of the 2-m band. Several of the more expensive and some of the later transceivers used dual-conversion receivers with IFs as low as 455 kHz. These units make the best candidates for FM conversion and, in general, are higher-quality units. Do not pay a premium for a transceiver with a VFO. For use with repeaters, the VFO frequency cannot possibly be set close enough or remain close enough for very long to effectively access a repeater.

Used FM transceivers are available at reasonable prices. The first FM transceivers available to the amateur were crystal-controlled units, and in some cases were part tube and part transistor. With the introduction of synthesized equipment, many crystal-controlled transceivers became available at low prices. Of course, no modification or modernization is required of the equipment if the crystal-controlled operation is acceptable. However, in many cases, these transceivers may be converted to synthesized by adding one of the several add-on synthesizers available in kit form. Occasionally, there was difficulty in interfacing the synthesizer to some transceivers. Be sure to check the availability of a compatible synthesizer before purchasing an older transceiver for later conversion to synthesized operation.

Transceivers with poor sensitivity can be improved by adding an RF preamplifier. Again, magazines and handbooks are good sources of schematics.

One very successful internal modification of the transceiver, if it suffered from poor selectivity, is to install an eight-pole or larger ceramic or crystal filter. Fortunately, most FM transceivers used standard IFs such as 10.7 and 455 kHz, which ensures a large selection of filters.

Another transceiver suitable for use on VHF bands is a commercial-band FM radio. These units are used in fire vehicles, taxicabs, police vehicles, and so on. They are of very high quality, and, when new, were very expensive. These transceivers usually operated on one or two crystal-controlled channels and may be easily converted for use in the amateur service. There are four types of units available. Low-band units, which cover 30 to 50 mHz, may be converted to 10 or 6 m. High-band units, usually covering 148 to 174 MHz, may be converted to 2 m. UHF units, which usually cover 450 to 470 MHz, may be converted to the 420- to 450-MHz amateur band. As they are rather new, 900-MHz units are not generally available for low prices. It is possible that units will appear in the surplus market in a few years and that some may be retuned to the 1-GHz amateur band. However, the method of construction of these units may make this modification difficult, if not impossible.

No modification of a commercial radio is necessary. It is necessary only to retune the unit to the appropriate amateur band. The tuning procedure must be performed exactly as outlined in the manufacturer's manual. A good signal generator, a VTVM, and a wattmeter are required for the tuneup operation. Some units may not tune low enough or high enough in frequency to reach the desired amateur band. If the amateur band is below the original frequency range of the transceiver, as would be the case of putting a high-band unit on 2m, it may be necessary to add small fixed capacitances across the tuned circuits that would not reach the new frequency. A capacitor value of 10 to 50 percent of the tuning capacitor should bring the tuned circuit down to the amateur band.

When putting a low-band unit on 6 m, the tuned circuits may not tune high enough to resonate on 6 m. The range may be extended by replacing the tuning capacitor with a unit of lower value, or by pulling the coil by the ends and "spreading" the turns.

Restoring and modernizing older equipment is not necessarily an activity limited to amateurs on a tight budget, but can be an educational and interesting experience for the more experienced amateur. Purchasing an old transmitter and receiver, restoring it, and using the equipment as a spare station can bring all the advantages of the spare station without great expense.

7 HOME CONSTRUCTION OF AMATEUR RADIO EQUIPMENT

The radio amateur has the unique advantage of being able to build radio transmitters and receivers for use on the air without any type acceptance from the FCC, a privilege that no other radio service has. All transmitters, and some other accessories for most radio services, must be accepted by the FCC before the devices can be sold. Although the radio amateur may build any sort of transmitter, receiver, or station accessory and place that unit on the air without formal approval from the FCC, this does not relieve the homebuilder from compliance with FCC regulations concerning spurious emissions, power-output restrictions, bandwidth, and so on, as outlined in the FCC rules and regulations, part 97, subpart C.

Many amateurs who construct their own equipment do so for economic reasons, and others construct equipment to obtain a certain type of performance that is not commercially available. But all amateurs who construct their own equipment receive a practical experience that is not available any other way. The knowledge of electronic fundamentals obtained from doing cannot be duplicated by reading textbooks.

7.1 SELECTING A PROJECT

Before any building can take place, a suitable project must be selected. Perhaps the most important criterion for building a specific piece of equipment is the desire to own or need for that piece of equipment.

The most desired equipment may not be the most desirable to build. A new amateur may desire a single-sideband transceiver, but it may be beyond his or her capabilities. In this case, abilities must overrule desire. Other factors, such as cost and difficulty, must be carefully considered, for it would certainly be a tragedy to spend a great deal of time and expense to build a complicated unit and lack the knowledge and experience to troubleshoot effectively and repair the mistakes. Similarly, it would be a tragedy to start a project and have to drop it for lack of funds.

The first decision to make is: What would be useful for the builder or necessary for his or her radio station? This will certainly be a large list of potential projects for most radio amateurs. Make a list of all the desired projects and find at least one article on construction for each item on the list.

With the list and the articles, estimate the difficulty of each project. Be sure to read the entire text of the articles and to look at the photographs. Some circuits that appear simple from the schematic diagram are actually very difficult to build because of some other facet that requires careful tuning or precise mechanical construction. Make sure that you understand each operation thoroughly, because any operation that you do not understand may indicate that it is beyond your capabilities. Be sure to understand what kind of test equipment is required to place the unit into operation, as some circuits that look simple require sophisticated tuning procedures for proper operation. It is not undesirable to reach just slightly beyond your abilities and gain by the experience, but it is of no value to reach way beyond your abilities and gain nothing.

7.2 DETERMINING COSTS

The cost of a project is about the easiest consideration to determine. First, look for obviously expensive items, such as mechanical filters, variable capacitors, power transformers, and RF power transistors. Prices for these items may be found in distributors' catalogs. Many expensive items may be on hand or obtained at a ham flea market for a nominal cost, but unless the builder is prepared for a long search, cut-rate sources cannot be counted on. Many amateurs have a certain type of device they specialize in building, such as receivers or VHF equipment, and are always on the lookout for parts suitable to these projects. This produces a well-stocked "junk box" for building receivers or VHF equipment and develops a keen eye for important parts. Many years ago, the junk box could be dipped into for just about any type of project, but with modern technology, it would be best to collect junk for specific projects.

A complete amateur station any ham could be proud of; this combination operates at 200-W power input on the 40-m and 80-m bands.

The cost of expensive items are added to the cost of the "nickel-and-dime" items, such as resistors, which may be estimated from catalog prices, and a "fudge factor" of 10 percent or so added for a safety margin. For the beginning builder, the cost of the home-built equipment may be as high as that for a commercially available unit. The purpose of building is more for the enjoyment and educational value rather than the economic value. Experienced builders can produce equipment for less than a commercially available unit because of the use of junk-box and flea market parts.

Finally, what kind of test equipment and special tools are required to successfully align and troubleshoot the project? Many systems, although they look simple, require expensive and hard-to-find test equipment to align. Although the builder may not have the required test equipment, a local ham may have the shop and the equipment to do the job, as well as offer some assistance. It is a good idea to ask around to be sure that the required test equipment will be available once the project is built.

7.3 OBTAINING PARTS

Once the project is selected, the next step is to acquire the parts. It is not necessary to acquire all the parts before building, as some parts may be installed after the entire unit has been constructed. Be sure, however, that the missing parts will become available. There is nothing more frustrating than to have a complete piece of equipment with one part missing.

Obtaining parts has been in a state of flux over the history of radio and electronics. There was once a time when any well-equipped radio store could supply all the parts required for just about anything. Of course, there were considerably fewer part types in those days. As the radio world developed into the electronics world and the number of different parts increased, no one supplier could ever stock all the parts in use. Gone forever are the days when a builder could obtain all the parts he needed "over the counter."

Many parts are available from mail-order suppliers that advertise in the classified sections of amateur radio and electronics hobbiest magazines. Integrated circuits and transistors seem to be the mainstay of these suppliers, with many suppliers having catalogs that list literally hundreds of items available by mail order. It is a good idea to request catalogs from several mail-order suppliers before ordering parts. The prices tend to vary, and handling and shipping charges can be large, especially if the order is small. Careful splitting of the order between several suppliers can reduce the overall cost and ensure a continuing supply of catalogs. It is also desirable to order some spares, just in case some parts are destroyed.

The biggest economic advantage of home construction is realized when the most expensive parts are obtained surplus or used. Surplus suppliers advertise through catalogs and magazine ads. It is not desirable to purchase expensive surplus or used parts without first inspecting them. Purchasing used and surplus items from a local equipment store is safer and often cheaper when the expense of shipping the large items is taken into account. Buying from other amateurs at ham flea markets is a very economical source of the larger and more expensive parts and allows the buyer to inspect the items before they are purchased. Either way, buying used parts is a world of "buyer beware."

There are many precautions that the builder can take to ensure obtaining useful parts for a reasonable cost. First, know what it is you are looking for, that is, what they look like, who made them, and approximately what they cost when new. Carrying a distributor's catalog as a reference book at a flea market could cause a lot of inflated prices to come down without haggling.

Many parts may be tested with simple test equipment such as a VOM. A small VOM is available to most amateurs and can be easily carried while shopping for parts. The fundamentals of testing several components using the VOM will be outlined in the next few paragraphs.

Meters are excellent items to be purchased at a flea market, as they are very expensive when purchased new and are very popular items at most flea markets. Do not attempt to measure the internal resistance of a panel meter with the ohms scale of the VOM. This can cause an excessive amount of current to pass through the meter and cause the pointer to bend or damage the coil. Place the VOM in the lowest current scale, typically 50 μA, and attach the VOM leads to the meter terminals. The polarity of the leads is not important for this test. Hold the panel meter and rock it back and forth so that the pointer oscillates. The oscillation should be smooth without sticking and the VOM meter should deflect slightly. This test is valid for low-current meters and basically any meter movement that has no internal shunts or multiplier resistors.

Capacitors, especially large electrolytic capacitors, may be checked with the VOM ohms scale. If the meter probes are placed across the capacitor while the meter is in the ohms position, the initial reading should be a very low resistance and will increase steadily. The smaller the capacitor, the quicker the rise in resistance reading. Depending on the meter, this method is not dependable for capacitors smaller than 10 μF. It is valid for locating shorted capacitors of any value or voltage. Since the behavior of the meter depends on the meter type, it would be a good idea to experiment with known good capacitors before measuring unknown units.

The VOM may also be used to evaluate transistors and diodes. Diodes may be checked by measuring the resistance with the probes connected in one direction and repeating the measurement with the probes reversed. The resistance reading should be very low in one direction and nearly infinite in the opposite direction. The resistance reading in the low-resistance direction will depend on the diode material. Usually, germanium diodes show a lower resistance than silicon, and Schottky diodes show a resistance lower than either silicon or germanium junction diodes. Since the direction of the meter probes and the actual resistance shown depends on the meter circuit, it would be wise to experiment with known good diodes before using the meter to evaluate unknown diodes.

The junctions of transistors may be checked using the same methods as for evaluating diodes. The base-to-emitter junction and the base-to-collector junction each will appear to the VOM as diodes. Again, it is a good idea to experiment with known good transistors before evaluating unknown transistors.

Resistors may be simply checked for their proper resistance value, while a visual check for discolored finish will reveal a history of overstressed parts.

Variable capacitors should show an open circuit while they are rotated throughout their entire range. Any shorting indicates a defective variable capacitor. If the plate separation is wide, the capacitor can probably be straightened out, but narrow-spaced capacitors, such as those used in broadcast-band radios, can be difficult, if not impossible, to repair.

In general, all parts should be inspected for indications of burning or discoloration due to overstress. Also look for signs of corrosion due to water or chemical contamination. Make sure that transformers, motors, and all things that look military operate from 60 Hz and not 400 Hz.

Industrial distributors, although a good source of parts, do not usually deal with individuals. This is because the cost of accounting required to fill small orders makes it difficult to deal with the public. Most industrial distributors will deal with individuals or radio clubs if the dollar amount of the order is large. These distributors are in the business of making a profit on electronic parts, and if the order is substantial, it will be honored. It is necessary that items be purchased in minimum lot sizes and that the order meet the minimum dollar amount. This usually means that resistors, for example, must be bought in multiples of 50 or 100 pieces, all of the same value. Substantial savings can be realized from quantity buying. It would be worthwhile for amateurs to organize buying coops or use an established radio club to take advantage of quantity buying from electronic distributors.

Most major magazines take all possible steps to assure the accuracy of the articles they print, but it is not unusual for an error or two to appear in an article, and the best source of information concerning errors, hints for building and obtaining parts, and other useful information is the author. Whenever writing to an author, be sure to include a self-addressed, stamped envelope for the return reply. In some cases, printed circuit boards are available, and it is almost always worthwhile investing in the circuit board rather than creating a new board. There will be cases where the boards available are very expensive or when circuit modifications are desired, and in thse cases it will be necessary to make your own boards.

Having now acquired the parts, the correct schematic, and other helpful information, the actual task of construction can proceed. It is a good idea to set aside a work area where the partially completed project can remain without being disturbed with all the required tools nearby and ready for use. Good lighting and electrical outlets for tools should also be available.

Before starting any work, consideration should be given to safety. Most electronic equipment construction requires light sheet metal work and electrical soldering. After construction, electrical measurements must be made for checkout and alignment. The metal work should be done with the proper tools and the proper techniques. Power tools should either be double-insulated or equipped with a three-wire grounded plug installed in a proper three-wire grounded outlet. Safety glasses should be worn when working with sheet metal or soldering. Clipping of the component leads is an especially dangerous operation, since the cutoff lead flies away with a very high velocity, which requires adequate eye protection. After the unit is constructed and electrical

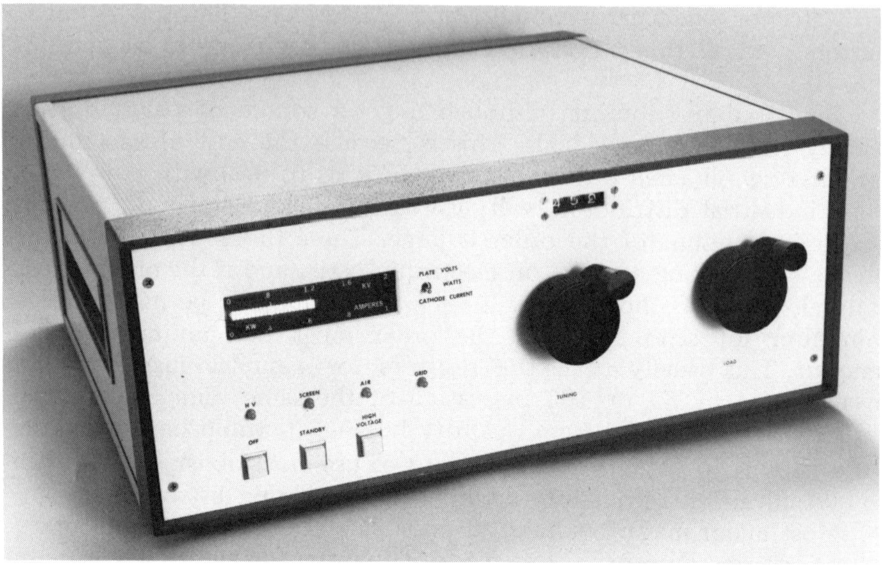

Example of a home constructed high-power linear amplifier. This unit features unique metering with a bar graph rather than a meter.

testing is carried out, extreme care must be taken to avoid electrical shock, especially if the unit has ac power supplies or if high voltage is present. Even if the unit has only low voltage present or operates from batteries, good safety habits should be followed. Developing bad habits while working with low-voltage equipment can jeopardize one's life when high-voltage equipment is being handled.

7.4 LAYING OUT PANELS AND CHASSIS

Unless the project under construction has a front-panel drawing that will be followed, the placement of parts on the front panel and the chassis, as well as the type of cabinet and so on, must be decided on an individual basis. The first decision to be made is what kind and size of enclosure will be used. There are many stock sizes of boxes and cabinets available from electronic distributors with varying prices, depending on style and size. The more attractive the case, the higher the price, and obviously, the larger the box, the higher the price. Most seasoned builders buy worthless electronic equipment for their cabinets and other hardware. Often an excellent cabinet can be salvaged from a piece of obsolete equipment and made to look like new with a can of spray paint and a new front panel. Whatever the source of the cabinet, it is important to choose a size that will house all the circuits, power transformers, connectors, and so on, without being overly large. If the equipment is to be portable, provision must be made for handles and extra feet on the side if the carrying handle is located on the side. Also, for rack-mounted equipment, handles can be fastened to the front panel to facilitate removing the equipment from the rack.

Building attractive, good-looking equipment requires a clean, well-thought-out appearance. This usually involves a functional layout with no traces of visible hardware.

Panel layouts may be evaluated by drawing scale-model sketches or by placing the knobs and other front-panel components on the blank, uncut panel or on a piece of heavy paper cut to size. The most important controls and indicators should be placed in the center of the panel and the less used controls placed away from the center, toward the sides and bottom. For example, when laying out a panel for a receiver, the main tuning dial should be placed in the center slightly toward the top of the panel, with other important controls, such as the audio and RF gain controls, nearby. Less used controls, such as the power switch and the headphone jack, can be mounted near the sides. Headphone and microphone connectors should always be mounted near the bottom of the panel so that the cords will not hang in front of the panel.

It is important when laying out any front panel that the placement of the controls and indicators be balanced. A well-balanced panel does not have a cluster of controls in one part of the panel while other areas have nothing. Panels should be balanced from side to side but not necessarily from top to bottom. There need not be as many controls on the top as on the bottom. Usually, the controls that require periodic manipulation are placed near the bottom, where they are convenient to the hand, while the top panel space is used for the dials and meters, where good visibility is required. Equipment that is to be carried as a portable unit or mounted under the dashboard of a car may require a completely different arrangement. A well-thought-out front panel should be aesthetically pleasing as well as functional.

Meters, dials, and readouts may be balanced with controls. For example, if a meter is located in the upper right-hand corner of a panel, it may be balanced by one or more controls, depending on the size of the meter, in a similar location on the left side of the panel.

The size of a knob should relate to the importance of the control and also the torque required to turn the shaft. An important function should not receive a small knob and be pushed off to the side of the panel. Also, large switches, such as band switches, that have heavy detent springs, should not receive small-diameter knobs. It cannot be overemphasized how important the appearance of the front-panel knobs is to the overall appearance of a piece of homemade equipment.

Many pieces of equipment have been constructed by the experienced amateur with parts practically from the junkyard that looked like new because of a brand-new set of knobs on a nicely painted panel with a good layout. It is also important that the knobs match. That does not imply that all the knobs must be identical but at least very similar. For example, if a black knob with a brushed aluminum insert is used, then all the knobs of varying sizes should have an aluminum insert.

Once a pleasing arrangement has been arrived at, it is necessary to determine if the arrangement is buildable. There are several things to watch for. If the front panel is to be attached to a chassis, some controls may be placed under the chassis, with others above the chassis. Be sure that there is sufficient clearance between the components and the chassis. Dials require a great deal of behind-the-panel area. Also, cabinets often encroach on the front-panel area. Be sure that the panel can be removed from the cabinet without hooking one of the panel-mounted controls; otherwise, it will be necessary to reposition some of the components to create a workable front-panel arrangement.

If a chassis is to be mounted behind the front panel, as is usually the case, there is a serious loss of flexibility in the placement of the front-panel components. It may be very difficult to create a balanced front panel given the restrictions of the chassis. This can be eased considerably by mounting the chassis away from the front panel by the use of standoffs, as shown in Figure 7-1. Components can be placed in practically any position on the front panel without fear of interference with the chassis. There are several details that must be considered carefully before a front-panel arrangement of this sort is used. First, the

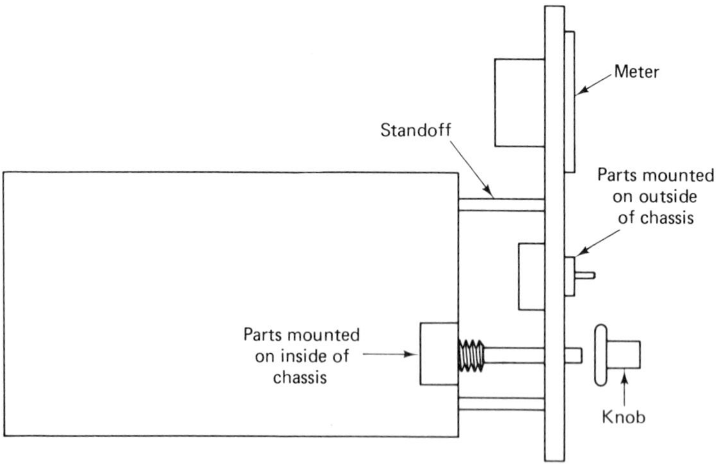

FIGURE 7-1. Method of mounting front panel with standoffs.

chassis, and hence its load, is supported by four small standoff spacers. The stress in this area is very large, especially if heavy items are to be mounted on the chassis. If the chassis must be large or carry a lot of weight, it would be desirable to include additional bracing. Second, there must be a method of attaching the standoff spacers without the hardware showing on the front panel.

Attractive equipment is made without any hardware, screws, nuts, or bolts visible from the front panel. There are several ways to achieve this with home construction. The first technique is to simply hide the mounting hardware. The most obvious place to hide a screw head is under a knob. This is not always possible because of interference problems with the control's shaft. There are, however, numerous applications where this can be done, such as attaching a subpanel. Another method is the use of blind fasteners. Several brands of blind studs can be used without special tools. These are usually headless screws that are pounded into soft aluminum with a hammer and a simple countersink-like tool and then sanded smooth. These fasteners, properly installed, are nearly invisible. If a screw head must show, it should at least be countersunk and painted.

One method of fastening the front panel without obtrusive hardware is not to hide it all but to use the hardware that is accepted. The nuts that hold toggle switches, phone plugs, meters, connectors, and so on, are acceptable on a front panel. In some cases, the hardware is of a decorative type which further enhances the appearance. Usually, these fasteners will have to be augmented with other fasteners.

A very professional method of removing mounting hardware from the front panel is to use a subpanel or so-called false panel. A panel slightly smaller than the front panel is mounted an inch or so behind the front panel and all the components are mounted on that panel. This method is effective only if there is a convenient method of securing the false panel without a lot of hardware showing. Shown here is a power amplifier that uses this type of construction. In this case, the false panel and the front panel both mount on the cabinet. Small securing screws are visible in the front panel, whereas the false panel contains numerous screws and nuts.

Finishing and decorating the front panel should be done carefully to assure a neat and professional job. The bare aluminum panel should be sanded with an orbital sander fitted with aluminum oxide production paper until all traces of high and low spots are removed. Deep scratches and blind fasteners that were set with a hammer may not sand out. If this is the case, these may be filled with an automotive-body-shop type of filler. This is often referred to as "spot putty" and is available in tubes. Before applying the putty, the panel must be washed to remove all traces of dust from the sanding operation. Once the panel

Power amplifier showing the technique of using a false panel for mounting parts. Compare this photograph to photo on page 259.

has thoroughly dried, apply the putty and allow it to completely fill the scratch or indentation. After allowing the putty to dry for several hours, the patched spot is carefully sanded with 320-grit carbide paper on a small wooden block. The panel is wiped with a rag soaked in common rubbing alcohol, and after drying, a coat or two of metal primer is applied. Some minor scratches may show up after the application of the primer, but usually the panel is now free of defects. The color coat is applied carefully, since any runs or sags will require that the entire panel be sanded and repainted and possibly refilled. It is best to check the spray paint on a piece of scrap and to apply several light coats rather than try to cover the entire panel with one coat. Once the color coat has thoroughly dried, the front-panel markings may be applied.

The only realistic way to apply the legends to a front panel is to use the dry-transfer letters available from most radio supply houses. These are legends on a plastic backing that literally "rub off" when pressed with a ball-point pen. Before applying the labels, be sure the paint is dry, and clean the painted panel with rubbing alcohol. This removes any traces of oils or dirt that may have accumulated during the painting process. Two important things to remember while working with dry-transfer labels: measure carefully and work cleanly. Once the labels are down, they are difficult to remove, so it is important that the positioning be accurate. Working with dirty tools and hands will invariably produce a smudge that will become a permanent part of the front panel. After the labels have been applied, the entire panel is sprayed with a clear coating of lacquer. It would be prudent to experiment with a piece of scrap metal before attempting a complete front panel. Some paints are not compatible, and the clear final coat may not be really clear at all, but affect the final color. It would be disastrous to spend hours preparing a panel only to have it ruined with the last coat of paint.

Although the front panel is laid out for appearance as well as function, the chassis layout is strictly functional. Two criteria apply: electrical reasons and mechanical reasons. The most important mechanical criterion in laying out a chassis is to place the heavy components where they will not damage the chassis or equipment. That usually involves placing the power transformers near the edges of the chassis and not in the middle of a large, unsupported panel. Also, it is important not to place heavy components at the end of a long cantilevered chassis. Heat sinks must be placed on the outside of the chassis if they are to work effectively. This usually implies the back apron of the chassis. Unless the equipment is to operate in an unusual environment, there are very few mechanical restrictions to the chassis layout.

Electrical requirements will dictate the majority of the chassis layout. First, the assemblies must be as close as possible to the controls and circuits that they interconnect, and remain as far as possible from those circuits that could cause interference. This means, for example, that an audio amplifier should be mounted as close as possible to the volume control and as far as possible from the power supply, which could cause ac hum pickup. In some cases it will be necessary to provide small aluminum boxes to individually mount some of the circuits to provide complete isolation. Usually, circuits that require complete shielding will be indicated on the schematic. In RF circuits, this could be practically everything, so that the chassis is nothing more than a shelf for small aluminum boxes. There are many standard sizes of utility aluminum cases from several different manufacturers suitable for subassembly shielding. There is also a lot of commonality in the sizes between the manufacturers.

Printed circuit boards can be mounted in several ways. First, a large hole, nearly the size of the printed circuit board but large enough to clear all the conductors on the PC board, can be cut in the chassis and the PC board screwed to the chassis. There are several disadvantages to this method. First, a large rectangular or irregular hole is difficult to cut into a chassis. Second, a large number of holes can weaken the chassis, and if the chassis is weakened and allowed to flex, the PC boards will suffer damage. On the plus side, it is possible to reach both sides of the PC board for troubleshooting or removing components without removing the board from the chassis. Another method, one that is preferred for home construction, is to mount the PC boards on short standoffs either on top or on the bottom or both sides of the chassis. Only four small holes have to be drilled into the chassis, rather than a large hole. Also, a certain amount of shielding may be obtained from the chassis. The major disadvantage is that the board must be removed whenever a component is changed.

The electrical wiring in a modern electronic circuit often involves nothing more than inserting components into a printed circuit board and soldering the components to the foil. In addition, PC boards must be interconnected and supplied with the proper operating voltages. Therefore, the modern electronic chassis should be a collection of printed circuit board or shielded boxes with several bundles of cables. Certain precautions must be taken or even this seemingly simple arrangement can cause trouble. It is a good idea to run the leads that carry 117 V ac separated from the rest of the circuits so that there is less likelihood that the 117 V will come in contact with any of the circuits. This will also help to reduce the effects of ac hum. Other wires that must be shielded or are unusually sensitive are usually noted on the schematic and should be treated accordingly. Different-color wires should be used for wiring to aid in checking and troubleshooting. The wires should be fastened to the chassis with cable clamps and be tied together with lacing cord or plastic cable ties. Wires that must pass through the chassis or any of the aluminum shielding boxes must do so with a connector, a feedthrough device, such as a feedthrough capacitor, or a rubber grommet.

7.5 WIRING WITHOUT PC BOARDS

Some circuits will not have a printed circuit board and will require wiring between the components. A perforated board with holes every 0.1 inch in both the vertical and horizontal axis is ideal for wiring integrated circuits, especially digital devices. The integrated-circuit sockets can be placed in the holes in the board and their leads wired together with hookup wire. For very large circuits, this method is not preferred because of the long time required to wire such a board. For making large digital circuits without the use of a printed circuit board, wire-wrap techniques could be used. Wire-wrap connections are made with small-gauge wire, usually No. 30, by stripping about 1 inch of insulation and wrapping the bare end plus a turn or two of the insulated portion around a gold-plated square post. No solder is used so that the wire-wrapped connection can be removed by simply unwrapping the connection. Wire-wrapping techniques are applicable to digital circuits and because of no shielding, are not useful for analog, or RF, circuits.

Making RF circuits without a PC board can be quite successful and easy. The most common method of manufacturing RF prototypes has no official title but involves supporting parts by their leads between a ground plane and feedthrough capacitors. If done with skill, this technique can produce stable RF circuits that are easy to build and repair but may not look attractive. Figure 7-2 shows a schematic of a single-stage RF amplifier, and Figure 7-3 shows the same circuit redrawn to include several feedthrough capacitors and to indicate the parts that will be on the reverse side of the double-sided copper board. Figure 7-4 shows this circuit constructed with a piece of double-sided copper clad board without any circuit etched into the surface. Many parts can be soldered to the copper-clad board without damage, which makes this a popular method of creating prototypes or one-of-a-kind circuits for RF frequencies, where appearance is not important.

There are two types of universal etched circuit boards available from manufacturers, completely etched and drilled, that may be used for creating different types of circuits. One type is the universal DIP board. This board has an etched and drilled pattern that accepts dual-in-line integrated circuits. This type of board is particularly useful for digital projects that require large numbers of integrated circuits. Another type of board has a large number of isolated squares of copper on

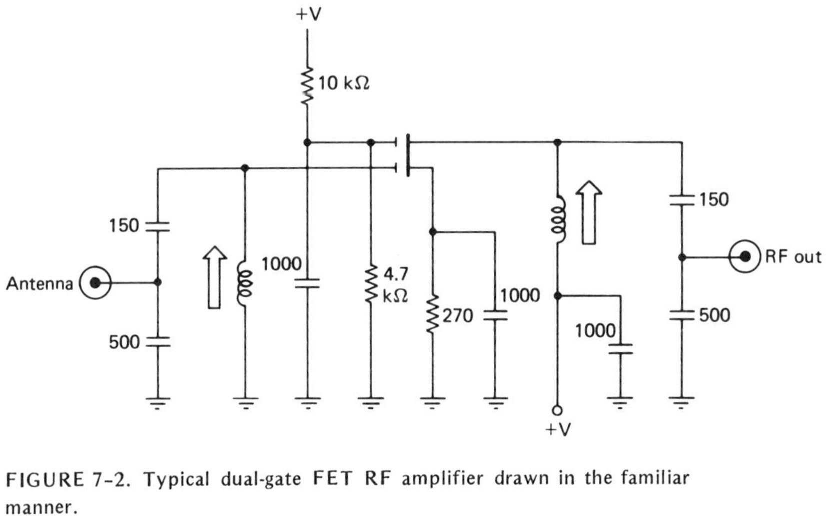

FIGURE 7-2. Typical dual-gate FET RF amplifier drawn in the familiar manner.

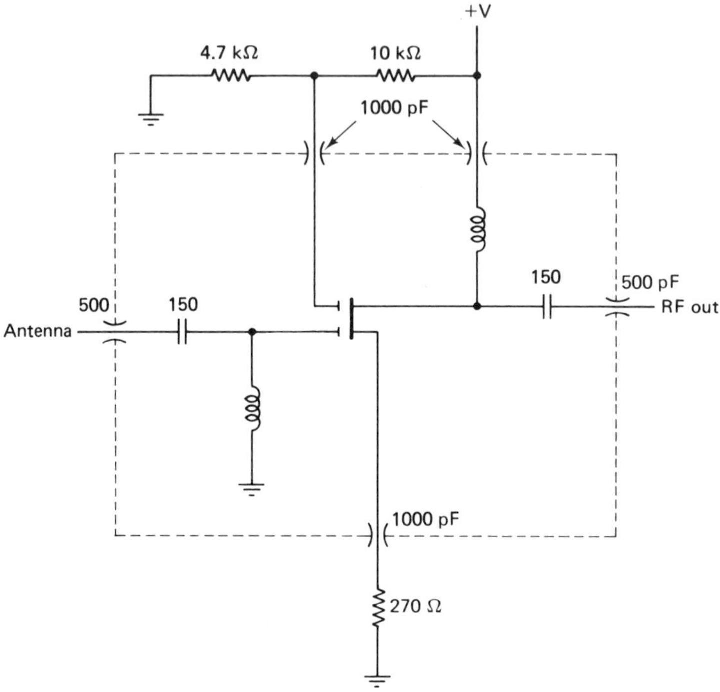

FIGURE 7-3. The same amplifier shown in Figure 7-2 but drawn to use feedthrough capacitors.

which circuits may be soldered and interconnected. This type of board is suitable for RF circuits and other small circuits. Because the board has no holes, it is usually inexpensive.

No electrical project can be successful without good soldering techniques. Many home-constructed equipment fails after a short period of use due to poor soldering, the most prevalent mistake being the cold

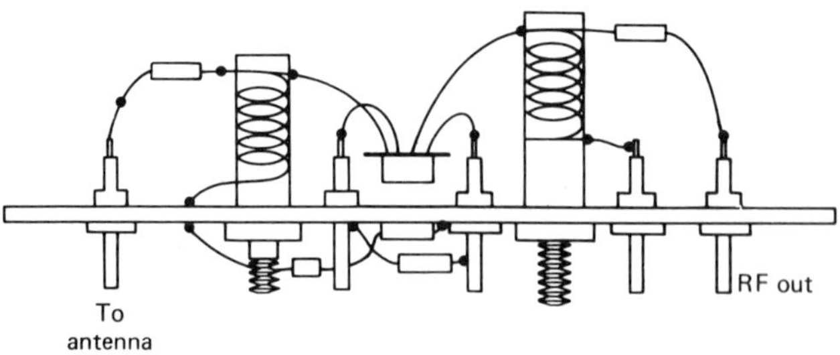

FIGURE 7-4. The RF amplifier shown in Figure 7-3 built on double-sided copper-clad board.

solder joint. This is a connection that was not brought to a high-enough temperature to properly flow the solder, or was disturbed while cooling.

The first step toward good soldering is to choose a good-quality soldering iron suitable for the task intended. For soldering printed circuit boards, a 25- to 50-W iron is sufficient, with a tip temperature of about 750° F. Most inexpensive soldering irons are not thermostatically controlled, and the variation in tip temperature could be 100° F or more. For most amateur work, a temperature-controlled soldering iron is not necessary. Most soldering irons have interchangeable tips that vary from 1/16 inch to 1/4 inch or more. The very small tips, although easy to use for PC board work, tend to burn up quickly and thus are uneconomical. Some tips are iron-coated and come pretinned. If treated carefully, these tips are very long lived. It is very important not to file the iron-clad tips. The iron cladding will be worn away and the resultant hole will decay very quickly.

As important as the soldering iron is the solder used. A good-quality rosin-core solder with 40 percent lead and 60 percent tin is recommended. Under no circumstances should acid-core solder or paste or liquid fluxes be used. Very thin solders make PC board wiring a little easier, while thick solders are excellent for soldering shields and large components. A good compromise is wire solder about 1/16 inch in diameter.

An important thing to remember about soldering: heat is applied to the work and the solder is melted on the work. Applying the heat is easier if the soldering iron is kept clean and freshly tinned. Wiping the iron on a wet sponge periodically will clear away any oxide that may have accumulated, and a fresh coat of solder on the tip will increase the heat-conducting capability of the iron to the work.

After the joint has heated up, touch the solder to the joint and allow the solder to melt freely around the joint. Do not apply excessive solder to the joint. If solder fills the area between the two wires, or bridges the entire gap between the PC board and the component lead, the joint has received sufficient solder.

A good solder joint appears shiny when it is made and may turn only slightly dull after it cools. A good solder joint never has jagged crystalline surfaces. Occasionally, some of the rosin flux will accumulate around the solder joint and may easily be removed with trichloroethylene.

Invariably during the course of construction, a joint will be soldered and, because of an error, will have to be removed. Using the proper tools can make a clean and neat removal job. Several tools are valuable for removing soldered joints; they are the solder sucker, solder wick, and various other desoldering tools.

The solder wick is most valuable for removing parts soldered to a printed circuit board. To use a solder wick, heat the joint until the solder has melted and touch the end of the copper wick to the joint. The copper will wick up the solder just like a piece of string will wick up water. It may be necessary to cut off the solder-filled end of the wick and repeat the procedure for larger joints. Usually, this will remove enough solder so that the component leads may be broken loose with a pair of pliers. For larger joints, the solder sucker is effective. This device has a spring-loaded plunger that literally vacuums up molten solder. The joint is heated and the tip of the solder sucker is placed near the joint. The plunger is released and the molten solder is drawn into the cylinder.

Several other unsoldering tools are available from suppliers. One simple but useful tool is a probe for unwinding leads that have been bent over the tie points. This tool is made of a material, such as aluminum, that does not take solder, so that the tool may be used on a molten joint without the tool becoming stuck fast.

After all the electrical wiring has been completed, the unit may be checked electrically. If the integrated circuits and transistors have sockets, certain electrical tests may be performed with an ohmmeter before the integrated circuits and transistors are installed. Measuring with an ohmmeter from one integrated circuit or transistor to another will verify that a connection does exist, but it does not verify that no extra connections are present.

After all the ohmmeter checks have been completed, insert the integrated circuits, or transistors, and apply the power. The first element that should be measured is the power supply current. An estimate should be made of the normal power supply current and compared to the actual current. If the current is excessive, immediately remove the power and check for defective chips or wiring errors. In many cases, this is not an easy task, especially where there are a large number of parts. One technique to determine where the excessive power is being consumed is to measure the temperature of the components trying to find parts that are hot. This is usually done with the fingers, but that is not recommended. A temperature-measuring probe or other device should be used; however, since most home constructors do not have such tools at their disposal, a careful check using the fingers is in order. Some integrated circuits operate warm to the touch, while others should not feel hot at all. CMOS logic, low-power Schottky logic, and low-power linear circuits should not feel hot at all. Power audio amplifiers, ECL logic, and voltage regulators may feel warm but never so hot that they cannot be touched.

After checking for excessive power consumption, functional checks may be made. A functional check implies that the equipment will be powered and checked to see just which functions operate. All the functions should be checked before any troubleshooting is attempted. It may be useful to know all the malfunctions before the unit is fixed. Commonality between the malfunctions can give greater insight into the source of the problem.

7.6 PRODUCING PRINTED CIRCUIT BOARDS

A printed circuit (PC) board can do much to enhance the appearance of a digital, analog, or low-frequency RF circuit and is often a must for high-frequency or high-power RF circuits. It is necessary for the experienced builder to develop a technique for the fabrication of PC boards. All PC board techniques are photochemical, with some differences in the way that the photo resist is applied to the copper-clad board. All PC boards start as a paper replica of the etched pattern of the finished circuit board. The method used to transfer this replica to the copper clad of the board for etching is the only difference in the two systems discussed. The first system is called the direct positive and is the easiest to use, but it does not produce the excellent high-resolution results of the negative method using reduction techniques.

Before a PC board is manufactured, it is necessary to produce a layout, often called the "artwork." Generation of the artwork starts with the schematic and use of rectangular grid paper, on which a plan for the PC board is laid out with a pencil. It is obvious that lines cannot cross one another, and it is often useful to redraw the schematic so that no two lines intersect. Often, this is not possible, so jumpers must be used for some runs. Figure 7-5 shows an example of a PC-board schematic. The schematic shown is a dual-gate FET RF amplifier similar to the type found on many high-frequency and VHF receivers. Figure

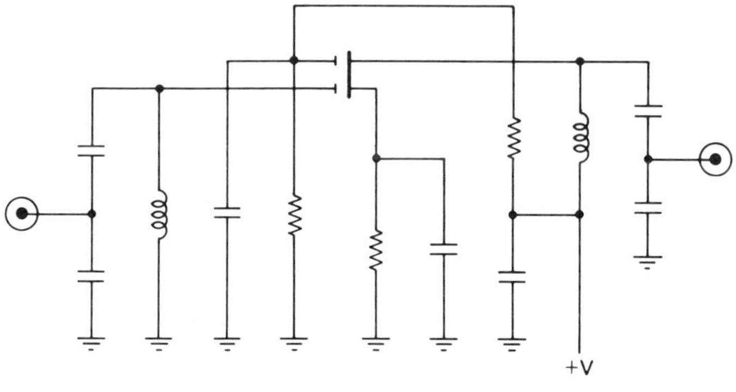

FIGURE 7-5. Normal schematic of a dual-gate RF amplifier.

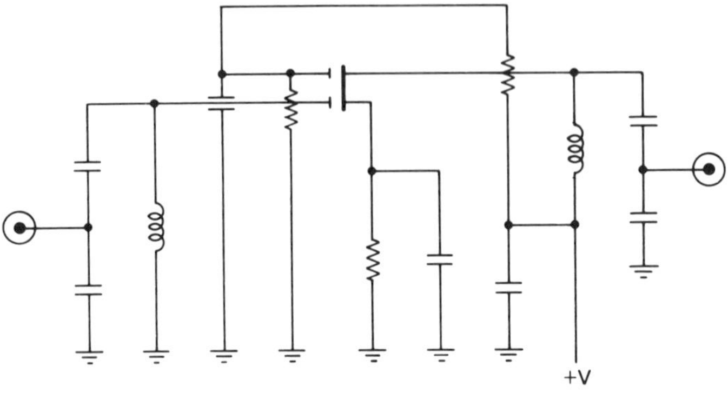

FIGURE 7-6. The same circuit shown in Figure 7-5 but redrawn so that no lines cross.

7-6 shows the same electrical circuit but redrawn so that no wires cross each other. Notice that electrical interconnects are allowed to pass through components, which usually would not appear in a properly drawn schematic. This schematic is now used for the basis in generating the artwork. The first step to generating the artwork is to draw the outlines of the components as viewed from the bottom. Artwork is always generated as if looking at the bottom of the circuit board, or as it is often called, the "circuit side," as opposed to the "component side," where most of the components reside. It is important to remember that many of the pin numbers for ICs and transistors are given in data books from the top side, or just the opposite, as viewed from the bottom of the circuit board. That is, to lay out a PC board, be sure to reverse the direction of the pin numbers from that shown in the data books.

If the layout is to be used for the direct positive method, the layout will be full scale; that is, 1 inch of actual circuit dimensions will appear as 1 inch on the layout. For those circuits that will be made with photo-reduction techniques, the artwork will probably be produced at twice the actual size, so that each inch on the layout will represent ½ inch on the actual circuit board. For either technique it is desirable to use grid paper with a light-blue gridwork with 0.1-inch spacing between the lines. Beware of grid paper with eight blocks to the inch or 0.125 inch between the lines. The appearance of this grid is similar to the 0.1-inch type and can easily be mistaken for the correct paper. The light-blue grid allows the artwork to be copied without any danger of the lines being copied.

The first step is to do a layout with a pencil at full scale or twice scale. This allows the design to be changed and modified by simply erasing the lines. The actual artwork will be generated with the help of adhesive-backed stick-ons that are manufactured with many different shapes common in the construction of printed circuit boards. For the normal runs between transistors or integrated circuits, where there is not a large amount of current, use 0.1-inch-wide tape, before reduction, which reduces to 0.05 inch after a 2:1 reduction. There are also round doughnut-shaped pads for terminating leads to a component. These come in several different sizes, with 0.25 inch, before reduction, being a good size. A doughnut that is too small will be damaged when the holes are drilled, and too large a pad will just simply waste space. Stick-ons with all the pads required of an integrated circuit are also available and are an absolute necessity if integrated circuits will be used. Other shapes are standard, such as corners, transistor pads, and so on. Usually, these shapes can be made with standard doughnuts and 0.1-inch tape.

Of course, artwork for the direct-positive method done at full scale would use tapes and patterns that are half as large as those used in the photo-reduction method—and therein lies the inherent disadvantages to that method. It is difficult to work with small-scale layout aids and more difficult to achieve any accuracy. The important advantage is that no reducing camera is required for the manufacture of printed circuit boards.

Once the artwork has been created, it is necessary to photograph it and reduce it to half-size. This requires a device known as a copy camera and, when obtained commercially, is a very large and expensive device. However, for simple reductions of 2:1, an inexpensive camera may be constructed using surplus parts. The homemade copy camera is a simple box camera with no shutter or iris and has a fixed-focus lens that can be obtained surplus at very low cost. Figure 7-7 shows a diagram of such a camera. The dimensions of the box are dictated by the focal length of the lens. For example, assume that a surplus office copier lens with a focal length of 6 inches was obtained. Using the

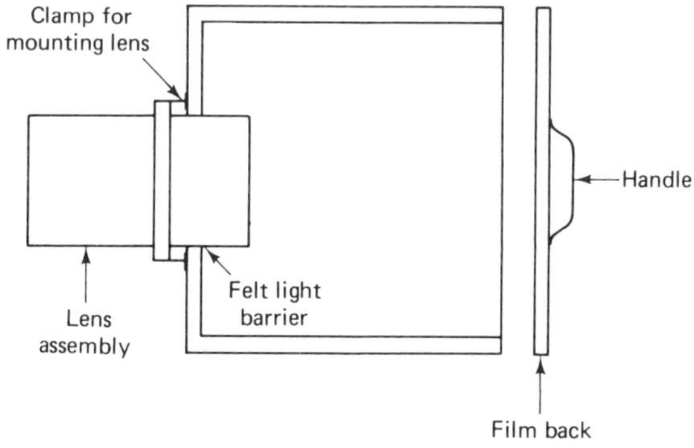

FIGURE 7-7. Simple copy camera for 2:1 reductions made from a surplus office copier lens.

This ugly duckling is a 2:1 reduction camera for fabricating printed circuit boards.

formula from a physics book, we obtain the reduction as the ratio D_1/D_2, which in this case equals 2, for a 2:1 reduction factor. (D_1 is the distance from the work to the lens, and D_2 is the distance of the lens to the film.) Since the lens will usually be a complex device several inches long, the actual distance may not be to the front of the glass or the middle of the lens tube or whatever, and it would be wise to assume that the distance is to the middle of the lens housing and make the lens mount an adjustable type so that minor adjustments may be made after the camera is constructed. A good choice for a lens is one from an office copier, which are plentiful on the surplus market and available for low cost. For the 6-inch-focal-length lens used in the example, the lensmaker's formula, again from a physics book, gives $1/D_1 + 1/D_2 = 1/\text{F.L.}$, where F.L. is the focal length. In this case, the focal length is 6 inches, and we have already put the restrictions on D_1 and D_2, so that $D_1/D_2 = 2$. Substituting, we obtain $3/D_2 = 1/\text{F.L.}$ or $D_2 = 3\text{F.L.}$ or, in this case, 18 inches. This distance has been dictated by the lens, but the other dimensions of the camera are left up to the builder. Because of possible imperfections in the lens, it would not be desirable to copy PC boards larger than about 6 inches by 6 inches. Two camera backs are required, a ground-glass back for setting up and focusing, and a film holder for making the actual photograph. Both of these backs are attached to the camera with magnetic fasteners, the type used to keep kitchen cabinets closed without moving parts. The camera must be totally light-tight, which can be checked by placing a small lamp inside the camera and looking around the outside in a darkened room. Any holes may be sealed with a wood filler or caulking compound. The entire camera, both inside and out, is painted with flat black enamel.

The camera back that holds the film should consist of a flat piece of wood with three small nails driven into the wood to mark the position of the film. It is easy to position the film in the camera back in a darkroom using the nails as guides. A camera back for focusing has a large hole cut in the center and is coverd with a piece of translucent Mylar grid. The grid markings, which are 0.1 inch apart and light blue in color, are available as drafting film from an art-supply store. A piece of grid vellum may be substituted but will not provide the accuracy or durability of Mylar film. It is important that the camera backs mount in exactly the same fashion and that the surface of the Mylar drafting film used for focusing be exactly the same distance from the lens as the plane of the film.

The camera has no shutter and must be kept in a darkened room whenever film is in the camera. A photographic darkroom would be an ideal place to use the simple camera, but any darkened room can be used. It is doubtful that any room in a normal house could be darkened suitably in the daytime, so that using the camera at night would be necessary. The artwork is fastened to a wall, or easel, so that it is parallel to the camera and about 3 feet from the front of the camera. A lamp, such as an ordinary desk lamp, is positioned in front of the camera so that the artwork is uniformly illuminated.

The first step in making a good-quality photograph is to focus the camera and adjust the reduction. For this task, the camera back with Mylar film is used. Focus the image on the Mylar film and position the camera so that the image is in the center of the viewing area. Using two points of known distance, such as the pins of an integrated circuit, determine if the dimensions have been reduced by a factor of 2. An eye loupe or magnifying glass would be helpful for accurate measurement. Professional high-accuracy artwork has a handy dimension reference. Two marks are placed on the original artwork and the distance is very carefully measured and marked on the artwork. When the work is reduced, the distance is carefully measured and set to be exactly one-half of the marked value.

If the image on the Mylar film is too large, move the camera away from the artwork, refocus, and measure the reference marks. If the image on the Mylar film is too small, the camera should be moved toward the artwork, refocused, and measured. Whenever the camera is moved, it must be refocused. This trial-and-error method should be continued until the reduction is exactly 2:1. For use at a later date, carefully measure the distance from the front of the camera and artwork and write this distance on the camera.

The film used for making printed circuits is a lithographic film that has extremely hard contrast, with no shades of gray whatsoever. For the film to retain this desirable characteristic, it must be developed with a special lithographic developer. Normal film developer can be used for the process but will not produce high-quality prints. Information concerning proper developing times, mixing, shelf life, and so on, is available from textbooks on industrial and graphic arts photography. This film and developer is available from well-equipped photo-supply stores at nominal cost.

Once the reduction has been established, remove the focusing back and tape a piece of film into the film back. Be sure to protect the back and the film from any extraneous light, which could ruin the picture. Place the film back on the camera and illuminate the artwork with the desk lamp. About 10 seconds is required for a typical exposure. The exact time may vary slightly and can be obtained by experiment. The amount of latitude for the lithographic film is considerable, so that it is not difficult to determine a usable time and produce good PC photos without waste. The film is removed from the camera back and developed according to the instructions supplied with the film or from a photographic data book. Once the film has been in the fixer for 5 minutes or more, it may be inspected for flaws. Two important characteristics must be inspected. First, the density of the darkened areas must be determined. Density is the ability of the darkened area to block visible and ultraviolet light. A sophisticated machine is usually used to determine this quantity in a photographic lab, but a good estimate can be obtained by guesswork. A good rule of thumb is to look directly into a 60- to 100-W bulb from a distance of about 3 feet through the darkened area of the film. If the outline of the bulb can be seen, the density is insufficient. The cause of insufficient density is underexposure, or insufficient development. A second picture should be taken with a longer exposure or longer development time, or both. The dark areas should also be inspected for voids and pin holes, which are usually caused by dirt on the film during exposure.

The nondarkened areas should be inspected for flaws. There should be no specks or patchy areas, which is an indicator of overexposure or overdevelopment.

For the beginner, it is an excellent idea to write down and record the temperatures, development, and exposure times that were used to produce good-quality lithographs. This information will be invaluable in future artwork reductions.

Even when all the temperatures and times are perfect, some small flaws can appear in the negative, which can be repaired. A sharp-pointed knife can be used to scrape away any darkened areas that are in a clear area of the film. An opaque paint may be obtained from a graphic-arts store to cover the pinholes and voids that appear in the darkened areas.

The image on the photolithograph, which is a negative, is of the correct size and will be used to transfer the circuit to the copper board for etching.

To photosensitize a copper-clad board, a photochemical, or photo resist, is applied to the copper surface. Good adhesion dictates that the copper surface be clean of oils, chemicals, and most important, copper oxide. Since the oxide is constantly forming, the boards must be cleaned immediately before they are photosensitized. Ordinary steel wool can be used, but a superior method is to use the plastic abrasive pads available from several manufacturers.

There have been several methods developed for coating the copper-clad boards with the photoresist to a uniform thickness. One method is to spin the board to be coated and apply the resist from the center of the spinning board. This requires special equipment and can be rather messy. A simple and good technique is to coat the boards by immersing them in a pan of resist and standing the boards on end to dry. A thick layer of resist will accumulate at the bottom of the board, rendering about $\frac{1}{4}$ inch unusable. Just make sure that the boards are larger than the circuit to ensure that sufficient board has been coated. Also, the boards are light-sensitive as soon as the photoresist has been applied. Therefore, the boards must be shielded from the light until they are to be used. The boards can be handled for a short period in a lighted room, but they should be shielded from such things as direct sunlight and long exposures to any intensity of light.

Before exposing the photosensitized PC boards to the negative, make sure that the negative, the PC board, and the work area and tools are free of dust. Any dust settling on the workpiece will appear as a void in the printed circuit pattern. Place the negative directly on the photosensitized board, being sure that the pattern is not reversed. This can be verified by comparing the negative with the original artwork. Place a piece of regular window glass over the negative and expose to ultraviolet light for about 5 minutes. An ordinary sunlamp suspended about 2 feet from the work should produce acceptable results. Be sure to avoid personal exposure to the ultraviolet rays while working with the sunlamp. A box could be built to shield the light from personnel, or a simpler method is to carry out the exposure in a closed room.

Once the copper board has been exposed, it must be developed and washed before the pattern can be inspected. Catastrophic problems will be visible, but most of the smaller and more prevalent problems will not be visible to the home builder. The only way to spot these small problems is to etch the board and assess the final PC board, Fortunately, many of the problems that may develop can be easily repaired.

Etching the PC board is a relatively simple job if done carefully. Ferric chloride is an excellent etchant and is relatively safe. The compound is not an acid and contact with the skin will not cause burns. Perhaps the biggest problem with the use of ferric chloride is its tendency to stain everything with which it comes into contact. Clothes, fingers, and utensils coming into contact with the ferric chloride will be stained with a mark that is difficult to remove. Use only glass and plastic trays and containers for the ferric chloride, to prevent any adverse reaction with metal. As the ferric chloride is used, it will increase its content of copper and lose its ability to etch printed circuit boards. This results in an increase in etch time. The etch time can be reduced by heating the ferric chloride with an infrared lamp. Small wisps of vapor from the etchant is not an indication of an overheated etch. However, a steady stream of vapor, or any sign of boiling, is a sure sign of an overheated etch, which should be reduced by moving the lamp farther away.

It is important to allow any precipitate formed at the surface of the board being etched to fall away from the board. If the products of etching were allowed to accumulate, they would cause localized slowing of the etch process and result in a nonuniform etch. The precipitate can be removed by agitating the ferric chloride to keep the solution moving. A much simpler and very effective method is to suspend the board near the top of the etchant container, using a piece of tape. The precipitate will simply fall to the bottom of the container.

The progress of the etch can be assessed by removing the board from the etchant and inspecting it. It is not unusual for certain areas to etch faster than others, so do not be alarmed by a board that seems to be etching faster in one area than in another. After the PC board is completely etched, allowing for specks and pinholes, it should be washed in running water for several minutes. After the board is dry, the surface should be cleaned with either a photoresist stripper or with a nonmetallic-wool-type pad. The board is now ready to be drilled.

Drilled holes in printed circuit boards are usually very small and numerous. The proper method for drilling printed circuit boards is with a very high speed drill and carbide bits. Unless a high-speed drill is available, carbide bits are not recommended. They are very expensive and brittle, which makes them break very easily. They have a very long life, cutting into the glass-reinforced boards if used properly. Regular steel high-speed bits will make acceptable holes in glass-epoxy boards but will have a shortened life span and cannot be resharpened. For the home construction of printed circuit boards, it is best to use high-speed bits for as long as they cut, and discard them after they will no longer cut through the glass effectively. It should be no problem to be able to drill at least 500 holes with a normal high-speed drill.

7.7 CONSTRUCTING HIGH-POWER AMPLIFIERS

Construction of high-power linear amplifiers has been popular with radio amateurs. Construction techniques for high-power equipment is somewhat different from the types of construction used for receivers and even those for lower-power transmitters. The first and obvious difference is that high-power linear amplifiers are vacuum-tube circuits and require high voltage, often forced air cooling and other devices unique to vacuum tubes.

The modern power amplifier is usually divided into two separate cabinets, one housing the power supply and the other housing the RF components. There are several good reasons for making this division. First, the power supply is usually a large cabinet and quite heavy. It is desirable to have this bulk located at a remote position, such as under the operating table, and its function controlled by remote control. The on/off function and standby ready switching can be done from a remote location, while any metering can be accomplished at the amplifier RF section. The RF section contains the vacuum tube or tubes and all the control circuits and the RF tuning circuits.

Great care must be taken when constructing RF power amplifiers to ensure the safety of those that may come in contact with the equipment. All wires carrying high voltage must be adequately insulated and inspected periodically for damage. All connectors carrying high voltage must be of the special recessed type specially made for high-voltage applications. Covers and doors to cabinets that contain high voltage must be interlocked so that removal of that protective cover will cause the power to be removed from the power supply. This applies to the RF section as well.

Protection from fire hazards is also required in high-voltage and high-power amplifiers. Adequate wire size should be used for primary leads, and adequate insulation should be used for leads carrying high voltage. The primary input voltage should be 220 V ac for powers above 500 W. A circuit breaker or fuse of the proper size should be used in the primary circuit for protection.

Parts placement is seldom critical in a high-voltage power supply except for mechanical considerations. Plate transformers and large filter capacitors are usually too heavy for standard radio chassis and require something more substantial. Since the terminals for most components used in high-voltage power supplies are accessible on top of the mounting surface, it is possible to mount all the power supply components on a flat plate. An aluminum plate about ¼ inch in thickness will support most power supplies wihout bending. Some sort of enclosure must be used to protect the circuits, and provisions must be made for connectors or terminal strips for interconnections with the line and the RF section.

Parts placement in the RF box, on the other hand, is very critical. Most home-constructed amplifiers are made from available parts, so that construction specifications are not being followed exactly as shown. Much of the parts placement is up to the individual builder. The first rule is to allow sufficient distance between any part used in the final tank circuit or on any high-voltage lead. Insufficient clearance will result in an arc or flashover and possible damage to the amplifiers or power supplies. Another important rule is to be sure there is sufficient isolation between the input and the output circuits, especially in common-cathode amplifiers with grid drive. If there is sufficient feedback to cause oscillation, excessive voltages and currents may exist and cause extensive damage to the amplifier as well as potentially harmful interference if the unit is connected to an antenna. A third rule is to make sure that there is sufficient cooling. Each tube type has different cooling requirements that must be followed, or damage to the tube will result. Some tubes can operate at full power without forced-air cooling, relying only on the natural convection set up by the elevated temperature of the tube. Other tubes must have cooling air even before their heaters are energized. Be sure to check the data sheet of the tube that will be used in the amplifier to determine the cooling requirements. Usually, ceramic tubes require that cooling air, flowing through the anode radiator and across the socket pins, be present at all times. The amount of air and backpressure are specified for each tube type. Glass tubes require cooling air mainly at the glass-to-metal seals, especially the plate. Often special heat-radiating plate cap connectors are required for complete cooling. Most tubes have special sockets, chimneys, and plate caps, which tend to be expensive but must be used for maximum tube life. The final requirement for good amplifier construction is adequate shielding. The amount of harmonic energy present in and around the tube in a higher-power final amplifier can cause a lot of interference to other radio services, especially television. If the tube and its immediate circuits are not properly shielded, this energy will be coupled to leads and wires and be radiated, as well as being radiated directly. The vacuum tube and the entire tank coil must be shielded by an enclosure that has no large holes or seams. Any removable covers must be fastened with sufficient screws, usually about one per inch or 1½ inches, to ensure complete isolation. If cooling air is directed through the shielding box, the seams should not only be tight to RF energy, but must not leak air. If commercially manufactured boxes are used for linear amplifier construction, it should be modified to increase the number of screws accordingly.

7.8 ASSURING COMPLIANCE WITH FCC REGULATIONS

FCC regulations concerning the amateur services are some of the most lenient of all rules regulating communication. There are, however, several FCC requirements that must be carefully checked for compliance.

The most difficult and the most important to check for is spectral purity. FCC regulations require that the harmonic energy be 40 dB down on the high-frequency bands and 60 dB down from the carrier on the VHF bands. The only sure method for checking for harmonic amplitude is to use a spectrum analyzer. Unfortunately, this equipment is not likely to be owned, even by the very well equipped amateur. The best method to use if a spectrum analyzer is not available is to build transmitter projects exactly the way they are shown in the instructions and to use all the proper low-pass filters. Since the advent of the FCC ruling requiring certain specific limits on harmonic radiation, amateur magazines have been very careful in assuring that all construction articles describing transmitters be checked for spectral purity.

Even if an article is followed carefully, there is a chance of excessive radiation. One way to obtain a rough check is to ask a local ham to listen for your harmonic as you transmit. Be sure that the amateur is close enough to hear a harmonic that would be equivalent to 1 mW or so, and that he or she is far enough away so that your primary signal will not overload the receiver. This method will not provide an actual figure in terms of decibels but will provide some assurance that the harmonic signal is attenuated. It is always a good idea to check for television interference as soon as a project is completed and before it is placed on the air. Discovering television interference before it bothers a neighbor is a good practice to follow.

Another FCC requirement to check for is the determination of the power input to the final RF amplifier. This will be an important parameter to measure when constructing a high-power amplifier. The FCC rules are explicit on the value of power that may be used on the amateur band. Before using any high-power amplifier, calculate its power input.

8 CONSTRUCTION PROJECTS: TRANSMITTERS AND RECEIVERS

8.1 INTRODUCTION

8.1.1 About the Projects

Since their beginnings, amateur-radio textbooks and handbooks have featured projects for home construction that have been very popular. Construction of equipment has always been an important part of amateur radio.

The projects presented herein are divided between two chapters. Chapter 8 describes transmitters and receivers and Chapter 9 discusses accessories for the amateur station. Projects are presented with several levels of difficulty. A station that is relatively easy to build would consist of the direct-conversion receiver and VFO transmitter from Chapter 8 used with the iambic keyer from Chapter 9. This combination will operate with about 10 W of output on any one amateur band. By adding the linear amplifier from Chapter 9, the power can be raised to over 100 W output, nearly the legal input power limit for novice licenses.

Another station having many more features can be constructed from the two-band superheterodyne receiver from Chapter 8 with its matching transmitter. The keyer and the linear amplifier would make a good station for someone beyond the novice license interested in operating a completely home constructed station. Aside from operating over 200 W input, this station features a crystal calibrator, S-meter, and crystal filter in the receiver, and a high-performance keyer with squeeze key operation for the transmitter.

For those amateurs who would rather build accessories for their station, Chapter 9 offers projects that can be added to practically any amateur station. A preselector for all frequencies above 21 MHz can put new life into an old receiver, while an FSK terminal unit may be the first step toward radioteletype operation. A crystal calibrator is an excellent first project, and the unit described has many desirable features.

Generally, the projects are for the lower-frequency amateur bands. This was done to make it easier to achieve success with the transmitters and receivers described. Amateur equipment operating below 10 mHz can be built and serviced using inexpensive test equipment, the type normally found in a television repair shop. Less-than-perfect construction practices will produce operable equipment more likely on the lower-frequency bands than at VHF. Since most of the equipment described is for only one or two bands, it would be desirable to choose a band that does not fade with declining sunspot numbers, so that the versatility of the equipment does not decline shortly after it has been built. An important reason for one- and two-band equipment is the difficulty in arranging complicated band switches. Band switching a large transmitter or receiver often requires complex switches that must be mounted rigidly for frequency stability. It is often easier, and in many cases less expensive, to make a single-band unit. The simple receiver described in Section 8.2 can be configured for any one of the amateur bands just by changing the value of the tuned circuits; similarly with the VFO transmitter in Section 8.3.

All the projects were designed to be attractive, and in some cases the metal work does not represent rock-bottom costs. It is possible that many of the projects could be built for less cost than shown if more austere cabinetry were used.

Finally, a word about the use of printed circuit boards and the general appearance of the electrical parts of the circuits. Since the actual equipment pictured was used to develop these projects, some of the interiors look as if they have been cut, spliced, and in general, mutilated. That is because they have been. Not every circuit shown started out in that fashion on the drawing board. Changes were made as the units were constructed, giving rise to their appearance. No printed circuit boards were manufactured because of the possible changes. There is no harm in creating printed circuit boards for any of the projects, although the non-PC techniques shown and outlined in Chapter 7 will produce satisfactory circuits if done according to the methods described.

Care and patience is the key to any successful project. After wiring, a thorough check is in order before applying power. This will reduce the possibility of destroying expensive components.

Before using the equipment for normal amateur radio use, some techniques used by manufacturers of high-reliability equipment, such as avionics, can increase the reliability of the equipment. One of the first tests these manufacturers perform is vibration. Obviously, no amateur is able to perform a serious vibration test. However, anyone can smack the sides of the equipment with a hand and turn the unit upside-down and give it a good shake. What will usually happen is that loose screws, cutoff component leads, and the like will be dislodged and fall out. Another check performed by manufacturers is accelerated aging or burn-in. Commercially, this would be a high-ambient-temperature environment, with the equipment operating at full duty cycle for about 72 hours. The amateur can perform a simple burn-in by allowing the equipment to operate on the corner of the workbench for a few days before it is put into service. Simple procedures such as these can make home-constructed equipment rival factory-made equipment.

8.1.2 About Parts Substitutions

As explained in Chapter 7, experienced builders can save considerable cost by using what is available to them, even though those parts may not be exactly what is called for on the schematic. There are some substitutions that the novice builder can make without difficulty.

Mica capacitors can be substituted for ceramic capacitors unless a temperature-compensating characteristic, such as N750, is specified. Any capacitor of the same type with a higher voltage rating can usually be used. Compression trimmer capacitors may be substituted for rotary units, and vice versa, unless the compression type is used specifically for its low inductance. Mylar capacitors may be substituted with polypropylene, polycarbonate, or polyester types. Tantalum and aluminum electrolytic capacitors may be interchanged unless the equipment is to be used at very cold temperatures.

Resistors of closer tolerance can always be used. Also, resistors with a higher power rating may be used except in high-frequency circuits. Never substitute a wire-wound resistor for a carbon resistor in an RF circuit at any frequency.

These are not the only substitutions permitted. There are many others that only experience will make available to the builder.

8.2 DIRECT-CONVERSION RECEIVER

A direct-conversion receiver (Figure 8–1) is especially suitable for a first project because of the relative simplicity of the circuit and good performance afforded by the design. It is also useful to the beginning amateur or those aspiring to become amateurs. This receiver is of a thoroughly modern design with special emphasis given to simplicity. Single-band operation was chosen to simplify the mechanical construction. Two-band operation would have involved a band switch, special shielding, and a more difficult calibration and alignment technique. In addition, the receiver is electronically tuned with a varactor diode to enhance the simplicity.

In spite of the simplicity, the receiver is a good-performing unit with a sensitivity of less than 1μV and a selectivity of less than 1 kHz.

The circuit consists of a dual-gate FET RF amplifier providing abou 20 dB of gain, feeding the active mixer. The dual-gate FET amplifier is very stable and has a good noise figure while providing good gain. The active mixer or product detector uses another dual-gate FET for an additional 15 to 20 dB of gain. The tuned circuits in the input of the RF amplifier, the output of the RF amplifier, and the mixer input provide about 100 kHz of band coverage without retuning. The VFO is set to cover about the same range in frequency by adjusting the voltage applied to the varactor diode and a trimmer capacitor in the oscillator. A 10-turn potentiometer provides a smooth backlash-free tuning system without fussing with gears or strings. The VFO feeds the product de-

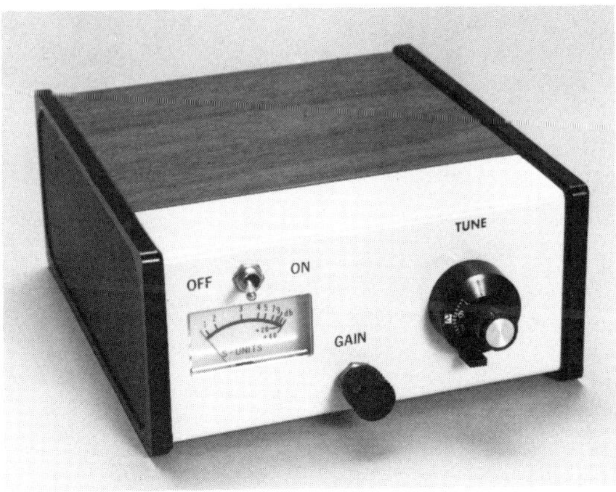

This direct-conversion receiver can be constructed for either the 40- or 80-m amateur band.

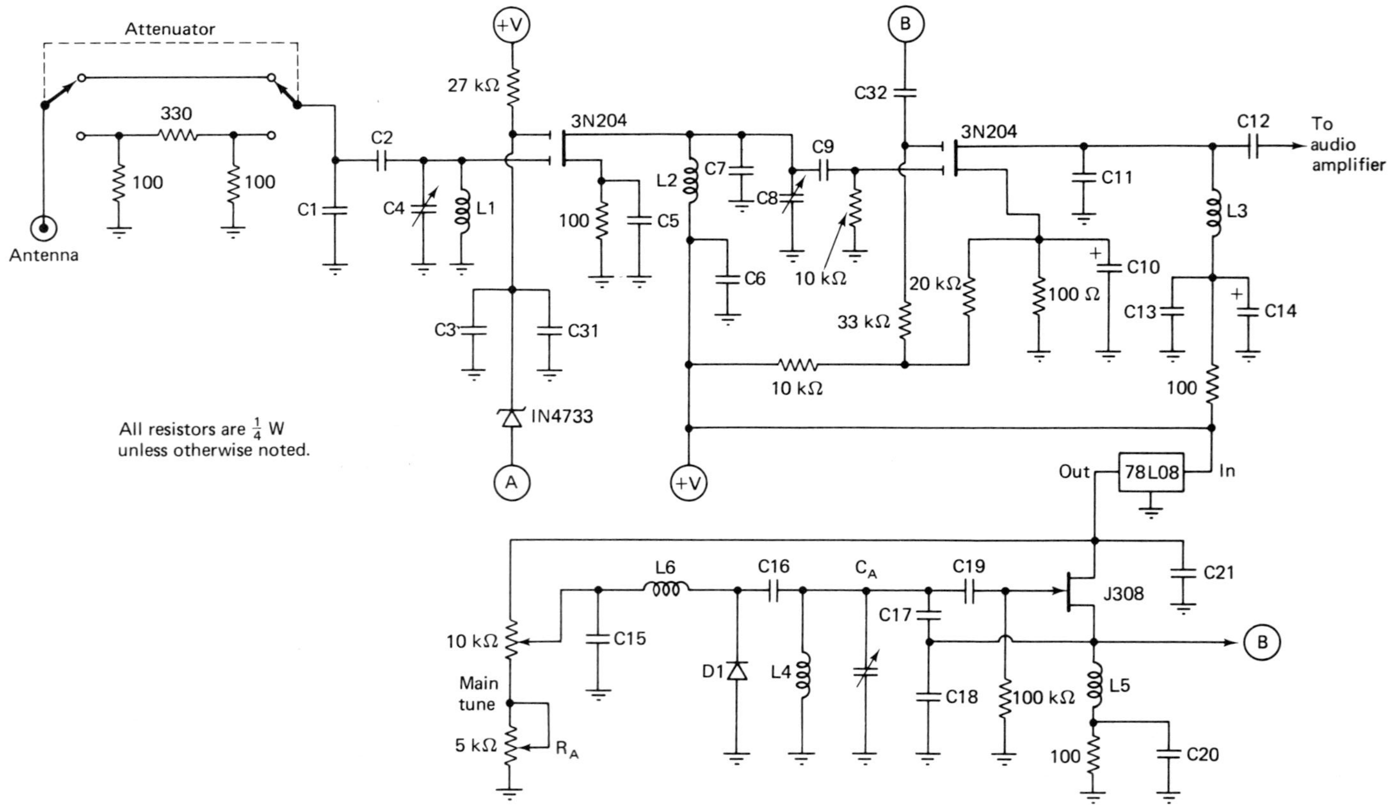

(a)

FIGURE 8–1. Schematic of the direct-conversion receiver.

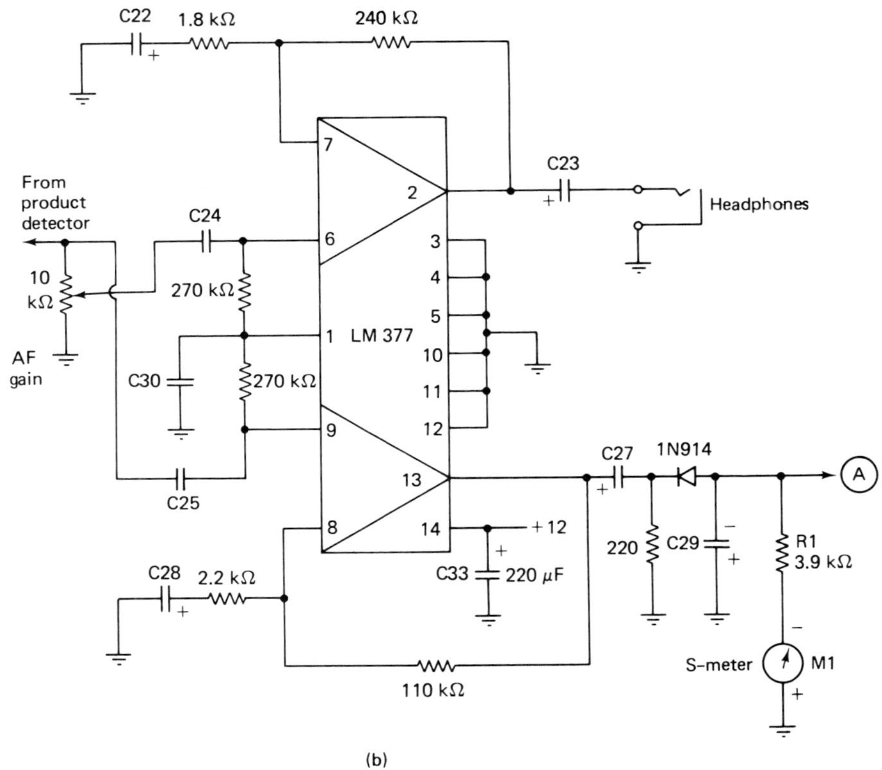

FIGURE 8-1. Continued.

Parts List for Direct Conversion Receiver

C3, C5, C6, C9, C13, C19, C20, C21	1000-pF ceramic, 50 V
C4, CA, C8	7-35-pF trimmer ceramic
C10, C14, C31	10-μF electrolytic, 20 V
C11	0.47 Mylar, 50 V
C15, C12	0.1-μF Mylar, 50 V
C16	15-pF mica, 50 V
C22, C28, C23, C27, C30	10-μF electrolytic, 20 V
C26, C29, C33	220-μF electrolytic, 20 V
C24, C25	0.1-μF Mylar, 50 V
C32	4.7-pF ceramic
D1	MV109 varactor diode
M1	S-meter 0–1 mA full scale; Other meters may be used by adjusting R1 by using the following formula: $R1 = 4/I_m$, where I_m is the meter full-scale current
R1	3.9-kΩ
R2	10-kΩ 10-turn potentiometer
L3	88-mH toroid inductor (see the text)
L5, L6	330-μH RF choke

40-meter components

L1	27-turns on T50–6 core
L2	26-turns on T50–6 core
L4	23-turns on T50–6 core
C1	1000-pF ceramic
C2, C7	150-pF mica, 50 V
C17, C18	330-pF mica, 50 V

80-meter components

L1	38-turns on T50–6 core
L2	37-turns on T50–6 core
L4	32-turns on T50–6 core
C1	2200-pF ceramic
C2, C7	330-pF mica, 50 V
C17, C18	620-pF mica, 50 V

FIGURE 8–1. Continued.

tector without a buffer amplifier by using a low-impedance point from the oscillator and feeding the second gate of the FET product detector.

The detected audio is removed from the drain of the product detector and filtered by a tuned circuit placed in the drain circuit. The audio amplifier, which provides the bulk of the receiver gain, is a low-noise stereo-power-amplifier integrated circuit designed to be used in inexpensive tape recorders and phonographs. One half of the chip is used to provide the audio amplifier and the other half is used for a very unique system for direct conversion—an AGC/S-meter. The second half of the audio amplifier is used to create a good audio-derived AGC system that uses the second gate of the RF amplifier for gain control. As is done in most AGC systems, the AGC voltage is used to power an S-meter for relative signal-strength indication. A zener diode in series with the gate of the RF amplifier and the AGC line ensures that the gate voltage will be about 5 V when no signal is being received. When a signal is being received, a negative voltage will be generated across C9 by the diode and will cause the voltage at the second gate of the RF amplifier to be lowered. In addition to the normal gain reduction supplied by the AGC system, a 20-dB pad is included at the input to the RF amplifier.

The entire selectivity for the receiver is obtained from a single tuned circuit in the drain of the product detector. A more modern approach would use an active filter, but the direct approach was chosen for simplicity.

The receiver can be constructed for practically any amateur band by simply changing the inductors and capacitors in the RF amplifier and the VFO. Values for the 40- and 80-m amateur bands are included in the schematic. Because of the difficulty of obtaining oscillator stability at the higher-frequency amateur bands, the desirability of the direct-conversion receiver becomes less at 20 m and higher. Twenty meters would give marginal performance, and higher frequencies would surely be a problem. The values shown provide about 100 kHz on either the 40- or 80-m amateur band. On the 80-m band it may be necessary to use two varactor diodes in parallel to obtain full 100-kHz coverage. If only one varactor diode is used, the tuning range may be only 50 kHz, but the VFO will be more stable. Coverage of the CW band only is desired because the receiver funtions best on CW only because of the sharp filter in the product detector.

Parts for the receiver are available from mail-order or electronic parts stores. The parts for the design were chosen very carefully to be available readily from well-known manufacturers. Most of the parts may be purchased new, but considerable savings can be had if some of the parts are obtained surplus, most notably the S-meter and the 10-turn potentiometer. Both items are popular at amateur flea markets and surplus stores for attractive prices. The 88-millihenry (mH) toroid is an item that must be purchased surplus. This toroid has been available to the amateur for so long from surplus outlets that it has become a standard part. These coils are used by telephone companies in large amounts and will be available for a long time to come. Magazine ads appealing to radioteletype enthusiasts should turn up a supply of these toroids.

Double-sided PC board is used to mount the majority of parts.

The entire receiver is constructed on double-sided copper-clad board using the techniques outlined in Chapter 7 using feedthrough capacitors and both sides of the copper-clad board. Constructing the receiver in this fashion causes no problems with oscillations or frequency stability. When building the receiver, be sure to observe the rules of good high-frequency construction and, in addition, be careful not to construct the oscillator near the antenna terminal. This allows energy from the oscillator to leak into the antenna circuit and be transmitted.

After the receiver has been constructed and the usual checking has been done, apply the power and listen to the audio from the speaker of headphones while the audio gain control is advanced. There should be some background noise without any loud oscillations or motorboating.

A calibrated signal generator or communications receiver is required for aligning the local oscillator. If the communications receiver technique is to be used, listen for the direct-conversion local oscillator signal with the receiver. It may be necessary to attach a wire to the antenna terminal of the communications receiver and place the wire near the direct-conversion receiver in order to get a strong signal. Set the 10-turn potentiometer to full clockwise, loosen the set screw on the potentiometer dial, and set the dial to read 95.0. This is assuming that the dial reads from 00.0 to 99.9, which most 10-turn dials do. Adjust C_A so that the frequency of the VFO is 7.095 mHz, as determined by the communications receiver.

Turn the 10-turn potentiometer to 00.0 and set the frequency to exactly 7.000 mHz with R_A. It may be necessary to set both end points more than once to achieve the best accuracy.

The RF amplifier and mixer tuned circuits are set by either using a signal generator or by peaking up on the air signals at 7.05 MHz. If the receiver is connected to a good antenna at night, the signals are usually so strong that something should be heard even if the amplifier and mixer are badly misaligned. Use the strongest signals first, then switch in the attenuator. Use weaker, but more stable signals, for the final tuning.

A signal generator may be used to align the receiver by simply feeding the signal into the antenna connector and using the signal generator to align the frequency and the RF and mixer tuned circuits.

The receiver is simple to operate and has sufficient sensitivity that something can be heard almost all the time. The audio gain control is used to set the listening volume while the stations are tuned in with the 10-turn potentiometer. If there seems to be a lot of very strong signals, as will almost always occur at night, switch on the 20-dB attenuator.

The receiver can be tuned for any 100-kHz portion of the band by simply setting the upper frequency of the oscillator with the 10-turn potentiometer fully clockwise and the lower frequency with the trimmer potentiometer. The RF and product detector tuned circuits are peaked at the center of each 100-kHz band. If the receiver was constructed for the 80-m band using the values shown, the technique is the same except that the receiver would be tuned for a 100-kHz portion of the 80-m band.

8.3 SINGLE-BAND VFO TRANSMITTER

The single-band VFO transmitter (Figure 8–2) is designed for CW operation only and is intended as a companion for the direct-conversion receiver described in Section 8.2. This transmitter is compact, battery-operable with good keying characteristics, and can provide about 8 W into the antenna from a 12-V supply.

This transmitter contains four stages: a high-stability VFO and buffer using junction FETs, which drives a high-gain power MOSFET intermediate amplifier which drives the 10-W power amplifier. VFO stability is assured by allowing the VFO to operate continuously and keying the intermediate amplifier. The power amplifier operates in a quasilinear mode, which is effectively a very low-bias current class B amplifier. This low-bias level would not allow for good distortion characteristics for single-sideband operation, but will give good keying characteristics for CW operation. The intermediate amplifier is a class B stage as well, but is keyed with a shaped keying voltage to generate the proper keying envelope.

The problems of VFOs that run continuously were covered in Chapter 2, and this transmitter uses a technique not covered in that chapter.

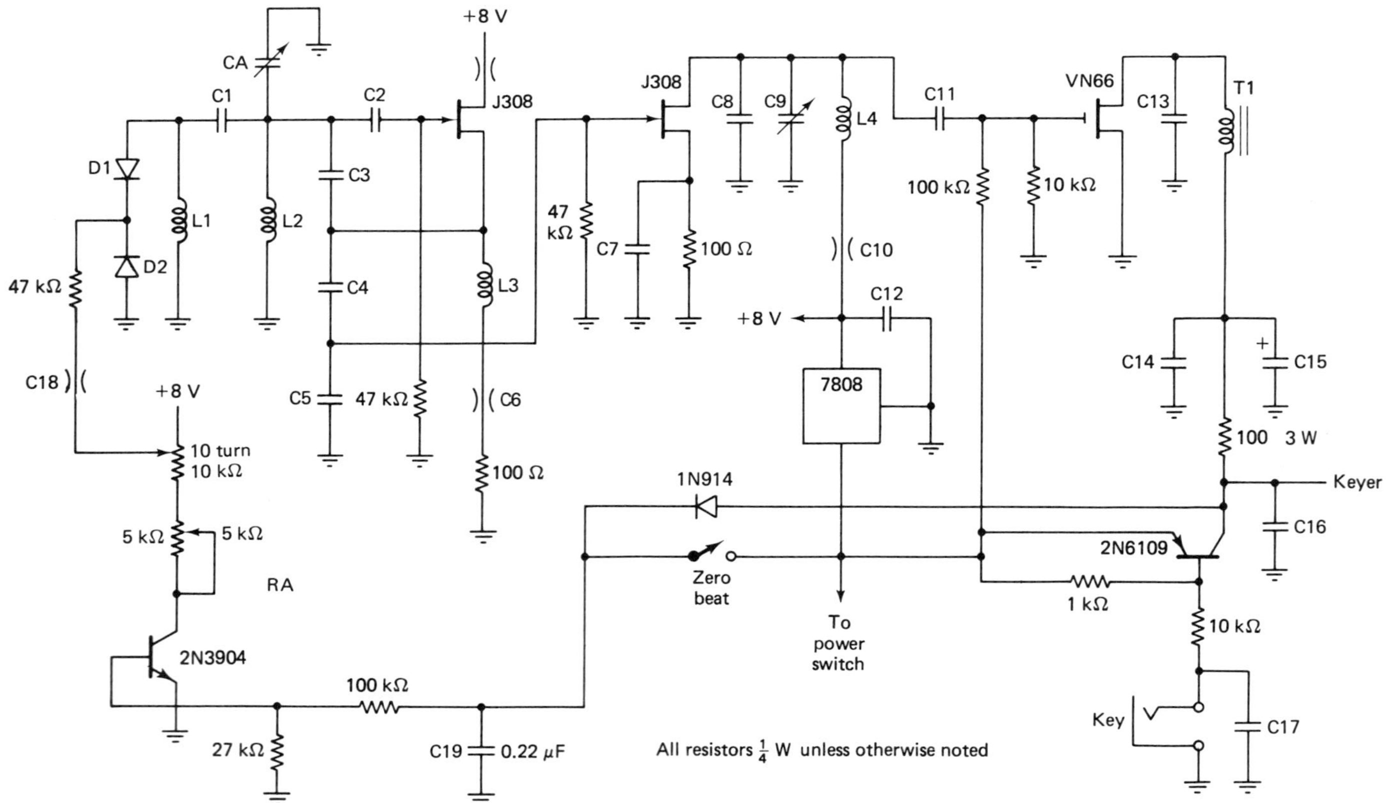

FIGURE 8-2. Schematic of the VFO transmitter.

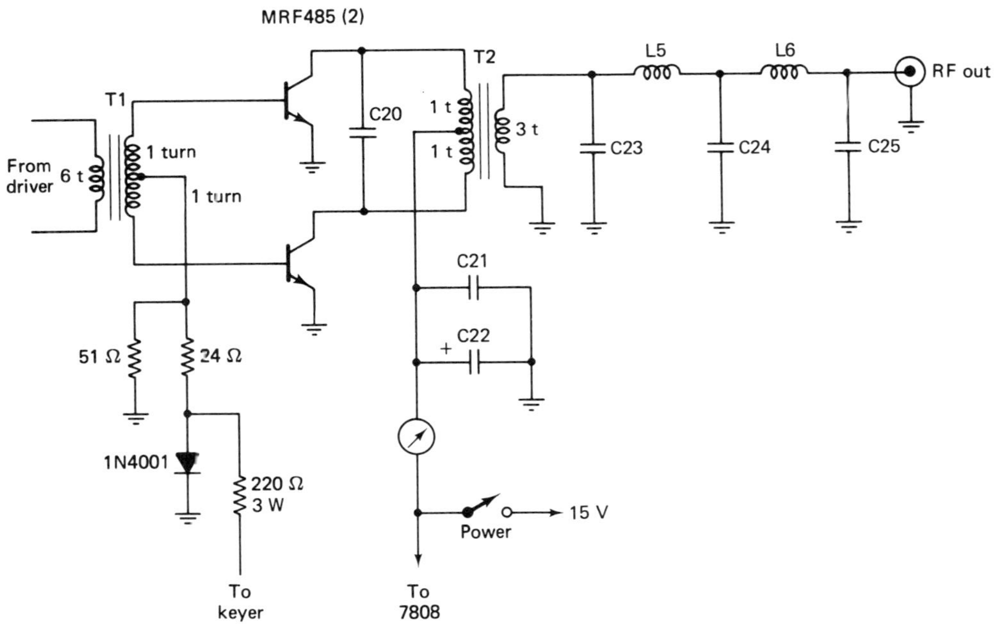

FIGURE 8–2. Continued.

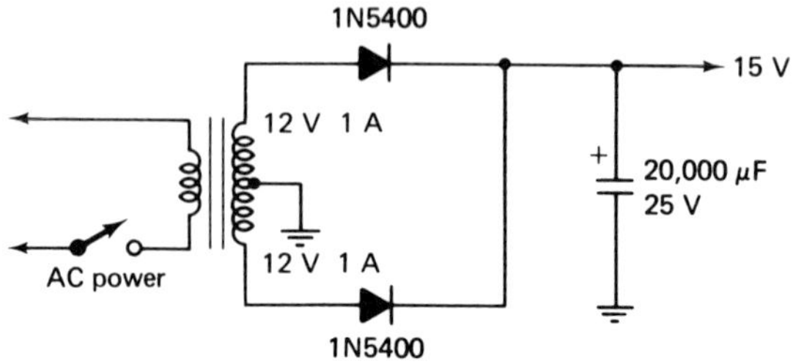

PARTS LIST

CA	7–35-pF trimmer
C1	47-pF ceramic, 50 V
C2	1000-pF ceramic, 50 V
C6, C10, C18, C26	1000-pF feedthrough, 100 V
C7, C14, C27	0.01-μF Mylar, 50 V
C8	500-pF mica, 50 V
C9	120-pF trimmer
C11	22-pF ceramic, 50 V
C12	0.1-μF Mylar, 50 V
C13	62-pF mica, 50 V
C15, C22	47-μF electrolytic, 25 V
C16, C17	0.01-μF Mylar, 50 V
C19	0.22-μF Mylar
C20	100-pF ceramic
C21	560-pF Underwood Electric Corp. uncased mica, 500 V
D1, D2	MV 109 varactor diode
L1, L3	330-μH RF choke

40-Meter Components

C3, C4	330-pF mica, 50 V
C5	1000-pF mica, 50 V
C23, C25	500-pF mica, 50 V
C24	820-pF mica, 50 V
L2	18-turns on T50–6 core
L4	14-turns on T50–6 core
L5, L6	19-turns on T50–6 core
C8	500-pF ceramic, 50 V

80-Meter Components

C3, C4	620-pF mica, 50 V
C5	220-pF mica, 50 V
C23, C25	1000-pF mica, 50 V
C24	1800-pF mica, 50 V
L2	25-turns on T50–6 core
L4	20-turns on T50–7 core
L5, L6	24-turns on T50–10 core
C8	1000-pF ceramic, 50 V

FIGURE 8–2. Continued.

The matching transmitter to the direct-conversion receiver. The "AM-FM" meter is a surplus item from a consumer item factory and gives testimony to the amateur's ability to use what is available.

In this transmitter, the VFO is shifted about 150 kHz up during periods of receive. Since the transmitter was designed to be a companion to the direct-conversion receiver, 150 kHz is sufficient frequency differential to ensure that the VFO frequency is beyond the range of the receiver. The frequency shift is accomplished rather simply by changing the tuning voltage on the varactor diode. Tuning the VFO electronically not only allows for the easy implementation of frequency shift, but simplifies the design of a suitable dial mechanism. A standard 10-turn potentiometer is used for tuning and the frequency is read out on a standard turns-counting dial.

The VFO drives a common-source JFET buffer amplifier from the source of the VFO transistor. The source is a lower impedance point than the gate and provides more isolation. The JFET buffer drives a common-source power MOSFET amplifier which provides over 20 dB of gain. The drain of the power MOSFET is connected to the primary of a wide-band transformer that feeds the push-pull MRF485 transistor amplifier. Some of the reactance of the driver transformer is tuned with capacitor C13. Capacitor C20 across the output transformer is for the purpose of reducing very high frequency components in the primary.

Since the driver and the final amplifier are strictly wide-band and contain no tuning, there is no attenuation to harmonics generated in those amplifiers. For this reason a five-pole low-pass filter must be used before the antenna to reduce the level of the harmonics to the legal limit. The push-pull amplifier affords some attenuation to even harmonics, especially the second, which is the most difficult to filter.

The VFO transmitter interior reveals the power amplifier and the harmonic filter mounted at the rear of the cabinet.

About 8 W of power output is available from the transmitter with a 12-V supply, while more power is available with higher power supply voltages. With the components shown, nearly 30 W is available with a 28-V supply. However, a larger heat sink should be used for the power output transistors at this power level.

The broad-band transformers are constructed from ferrite beads and brass tubing. The ferrite beads are not expensive but have been difficult to obtain. Manufacturers of ferrite products usually deal with custom-manufactured parts and have large minimum orders for their catalog items. A good source of beads is the Heath Company in Benton Harbor, Michigan. The part number is 475–17, and they are used in several Heath products. The brass tubing is used for fuel lines in model airplanes and is available from a well-stocked hobby store in 2-foot lengths. Two beads are placed on each brass tube as shown in Figure 8–3. The end plates are soldered to the ends of the brass tubes, which are also used for electrical connections. The high-impedance winding is 20-gauge hookup wire passed through the brass tubes.

When wiring any solid-state transmitter, it is very important that all leads be kept extremely short and that all bypass capacitors be wide-band. No capacitor can be low-impedance at all frequencies. Large capacitors offer low impedance at low frequencies, but because of

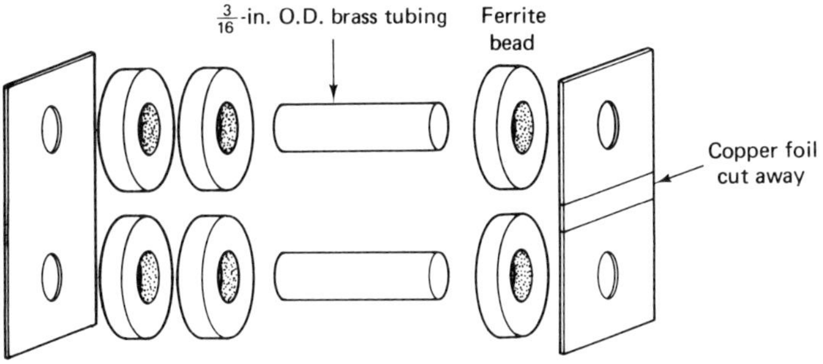

FIGURE 8-3. Method of construction of the wide-band transformer.

parasitic inductance, fail to provide effective bypassing at higher frequencies. On the other hand, small-value capacitors can offer low impedance at high frequencies, but because of the low capacity, do not offer much at low frequencies. A good bypass consists of three or more capacitors in parallel; a small-value ceramic capacitor for the high-frequency region, a Mylar capacitor for the medium-frequency region, and a large electrolytic capacitor for the low-frequency spectrum. Failure to use multiple capacitors can result in oscillations at a frequency where there is insufficient bypassing. Also, it is important that low-inductance leads be used for any ground connections. When building the VFO transmitter, be sure to follow exactly the layout shown in Figure 8-4.

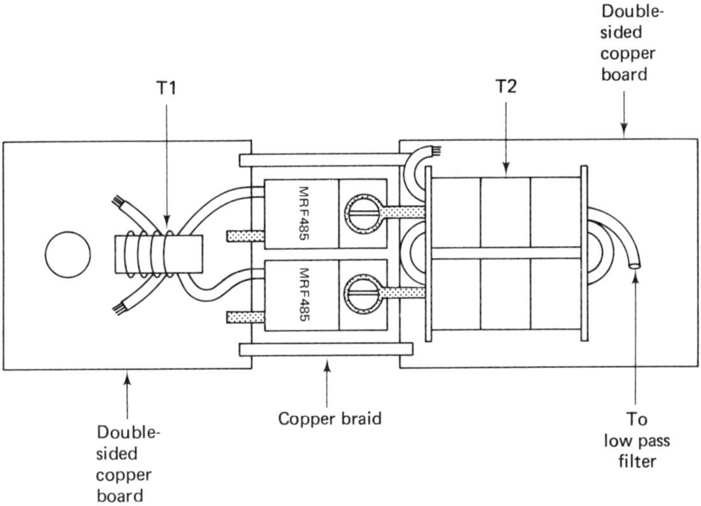

FIGURE 8-4. Layout of the power amplifier using the wide-band transformers.

During normal operation, the transmitter requires no tuning, but it must be adjusted before operation.

After construction and the usual checks for mistakes, apply the power. To check the VFO, measure the voltage at the gate of the VFO transistor with a voltmeter. If the VFO is operating, this voltage should be a couple of volts negative. Remove the knob from the 10-turn potentiometer and turn the potentiometer full clockwise. Reinstall the knob and set the dial to 95.0 (assuming that the dial reads from 00.0 to 99.9). Using the communications receiver, turn on the "spot" switch and listen for the VFO in the communications receiver. Adjust C_A so that the frequency is exactly 7.095 MHz. Set the 10-turn dial to 00.0 and adjust RA so that the frequency is 7.000 MHz.

Press the key; the transmitter should draw no more than about 1 A. If the current is less than 1 A, adjust C9 for maximum collector current or maximum output power as read on a wattmeter. The current should not exceed 1 A at 14 V. Using the ac power supply described, the power output will be about 10 W. Using a 12-V source, such as a battery, somewhat less power output will be obtained.

When operating the transmitter, maximum transistor life and power output will be obtained when the antenna is perfectly matched into 50 Ω. Most dipole antennas that are cut to the bottom portion of the 40-m band will have a low SWR and will allow the transmitter to provide maximum power output. If the antenna was cut for the phone portion or if there are any problems with the antenna or feedline, the SWR will prevent the transmitter from producing maximum power output. An antenna tuner will provide a method of matching the antenna to the transmitter as well as providing the necessary antenna changeover relay for the complete station.

8.4 SIMPLE SUPERHETERODYNE RECEIVER

Here is a superheterodyne receiver that uses only three integrated circuits, three transistors, and covers the CW portions of the 80- and 40-m amateur bands. The sensitivity of this receiver is better than 1 μV with excellent selectivity from a crystal filter and a good AGC system. Two additional features, an S-meter and a crystal calibrator, may be added as accessories for enhanced performance.

The basic design uses an old idea popular during the mid-1950s. A common local oscillator is used for both the 40- and 80-m bands. This is made possible by selecting the IF such that the 40-m band is the image frequency of the 80-m band. Therefore, changing bands involves switching only the front-end filter. This simplifies the receiver and

A wooden cabinet gives the superheterodyne receiver an elegant appearance.

enhances the frequency stability by not having any switching in the local oscillator circuit. Furthermore, fixed-tuned front-end filters are used for each band, to eliminate any tracking problems. The IF is chosen so that the image frequency of 3.5 MHz, the low end of 80 m, is 7.0 MHz, the low end of 40 m. Since the image is removed from the desired frequency by twice the IF, the intermediate frequency for the 40/80-m receiver is (7.5 - 3.5)/2 = 1.75 MHz. Therefore, the receiver tunes from 3.5 to 3.75 MHz and 7.0 to 7.25 MHz, and the dial is marked from 0 to 250. It is necessary to add 3500 kHz to the dial to determine the frequency on the 80-m band and to add 7000 kHz to the dial for 40 m. For example, if the bandswitch were set in the 40-m position and the dial read 85, the frequency is 7.085 MHz. If the dial were set to 150 and the bandswitch was in the 80-m position, the frequency would be 3.650 MHz.

The simple superheterodyne is a single-conversion receiver using an active balanced mixer for the front end and a single transistor plus one IC IF amplifier. Good intermodulation performance is obtained by eliminating the RF amplifier, often found in medium-performance receivers. The audio-frequency gain is obtained from an integrated-circuit audio amplifier that is capable of driving a small speaker or a pair of headphones. The beat-frequency oscillator and the local oscillator are field-effect transistors for high stability. The product detector is a part of the IF amplifier integrated circuit and is fed with drive from the beat-frequency oscillator. Automatic gain control is also a part of the LM373 IF amplifier. The S-meter is connected to the AGC control voltage line.

Bandswitching is accomplished by switching in the appropriate input bandpass filter with relays. It would appear that the use of relays is an unnecessary complication and that a simple switch could accomplish the task. Several important advantages are obtained from the use of the two relays. First, the isolation between the input and output of the filter may be made very great by locating the relays several inches apart and placing a shield between the input and output circuits. Second, a simple bandswitch is required, which allows it to be placed in a convenient spot on the front panel. Finally, lightning protection for the receiver when it is off, and TR switching, are by-products of the relays. Figure 8–5 shows the arrangement of the relays in switching the front-end filters. K1 switches the input and K2 switches the output. For 80-m operation, K1 is not energized but K2 is. This routes the antenna to the input of the 80-m filter while routing the output of the 80-m filter to the mixer. Both the input and output of the 40-m filter are grounded. This grounding is necessary to prevent energy from the antenna leaking into the 40-m filter and leaking into the mixer, causing spurious signals. On the 40-m band, the relay states are reversed, with K1 energized and K2 nonenergized. This routes the antenna to the input of the 40-m filter, the mixer to the output of the 40-m filter, and grounds both the input and output of the 80-m filter. Notice what will happen if the receiver is either turned off or the jumper marked "to TR switch" is removed. Both relays will be nonenergized. The antenna will be connected to the input of the 80-m filter while the output is grounded. The mixer is connected to the output of the 40-m filter while the input of the 40-m filter is grounded. This effectively isolates the mixer from the antenna, thus preventing damage during static discharges and during periods of transmit.

Signals passed by the input filter are fed to the active balanced mixer, which offers isolation of the local oscillator to the antenna and provides more than 20 dB of gain. The balanced output circuit of the mixer is used to good advantage in the crystal filter. An out-of-phase signal is required in a single-crystal filter to null out the effects of the crystal holder capacitance. On old receivers, an adjustment was provided to adjust this null accurately and was usually labeled "crystal phasing." The relatively high impedance of the mixer output is transformed down to a lower impedance that is more suitable for a crystal filter.

A FET Clapp oscillator covering the range 5.25 to 5.5 MHz is used for a local oscillator, which covers from 3.5 to 3.75 MHz on the 80-m

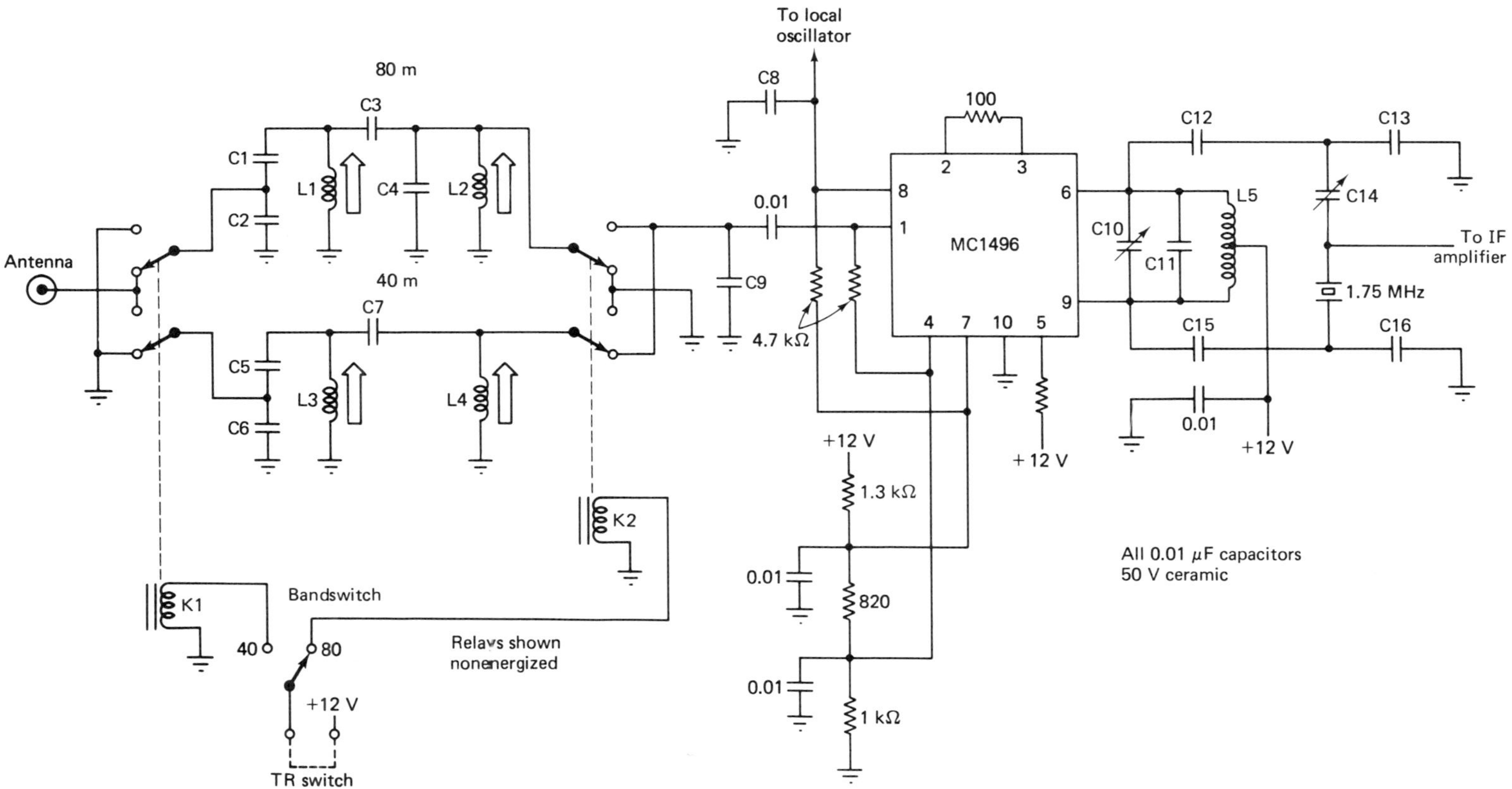

FIGURE 8-5. Schematic of the simple superheterodyne.

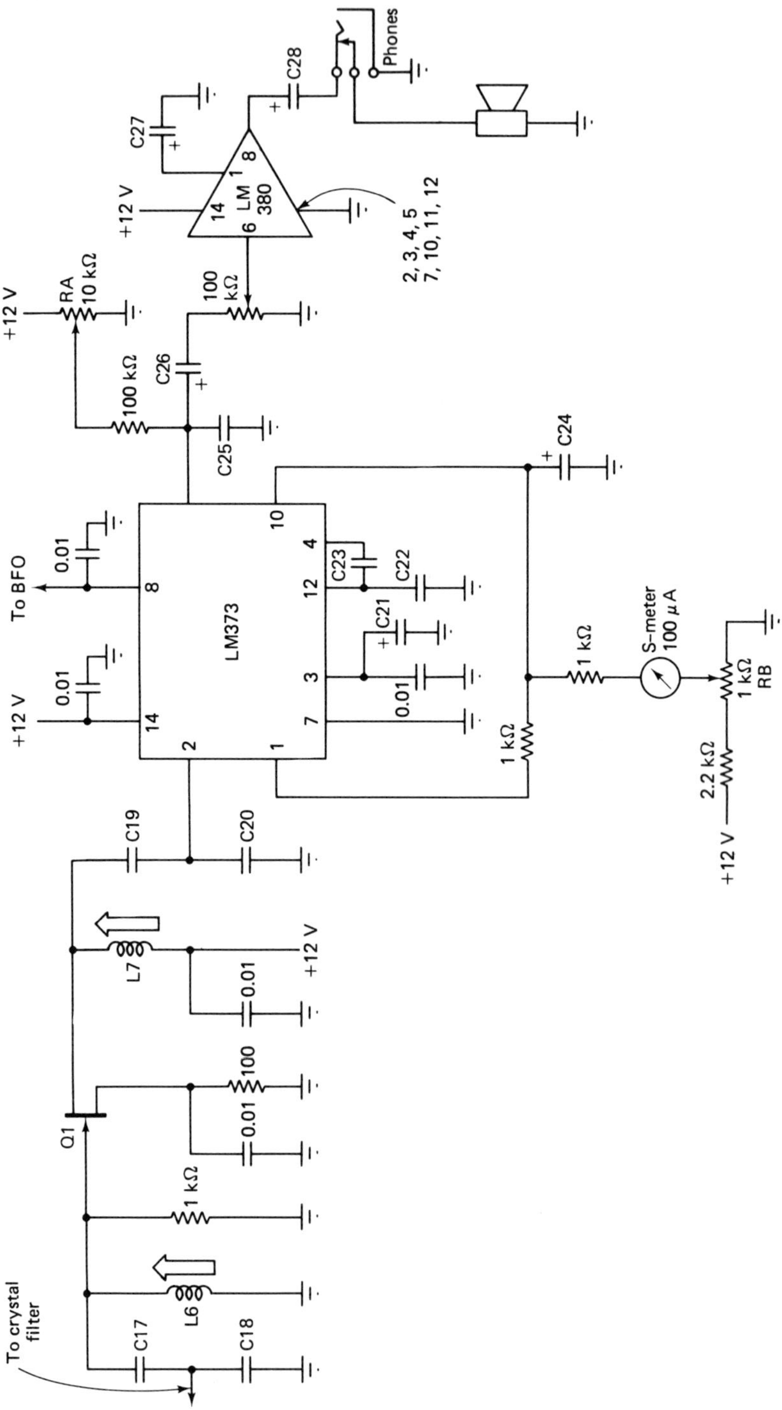

FIGURE 8-5. Continued.

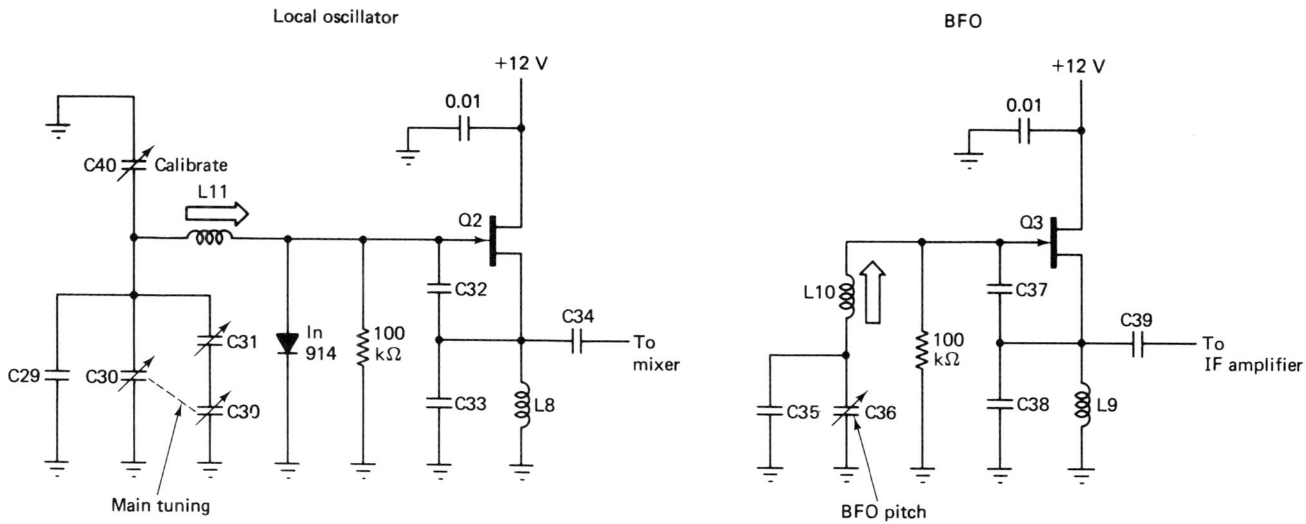

FIGURE 8-5. Continued.

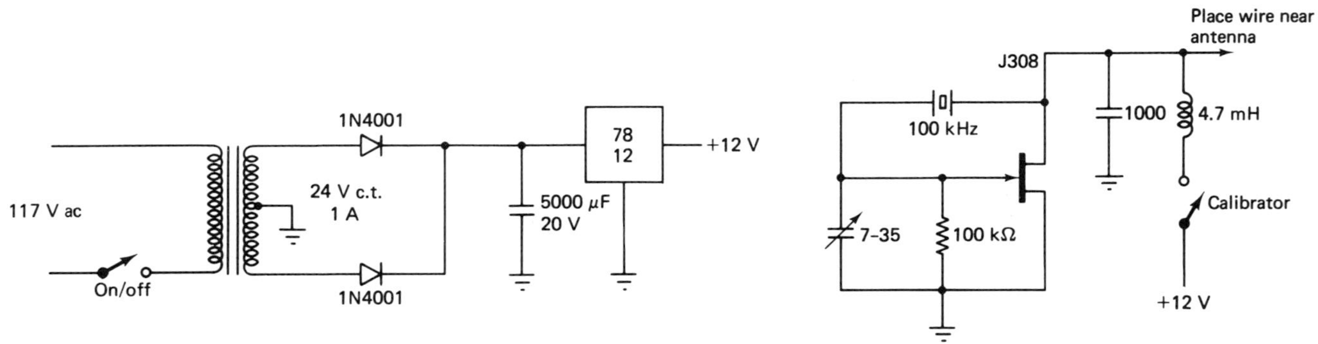

Power supply

FIGURE 8–5. Continued.

PARTS LIST

C1, C11, C17, C19, C35	330-pF mica, 50 V
C2	1000-pF mica, 50 V
C3	33-pF mica, 50 V
C4	150-pF mica, 50 V
C5, C8, C39, C34	220-pF mica, 50 V
C6	680-pF mica 50 V
C7	15-pF mica, 50 V
C9	180-pF mica, 50 V
C10	7–35-pF ceramic trimmer
C12, C15	1500-pF ceramic, 100 V
C13, C16, C37, C38	2200-pF ceramic, 100 V
C21, C26	10-μF tantalum, 20 V
C22	500-pF mica, 50 V
C23	0.003-μF Mylar, 50 V
C24	27-μF tantalum, 20 V
C25	0.05-μF Mylar, 50 V
C28	220-μF tantalum, 20 V
C29	75-pF mica, 50 V
C30	50-pF per section
C32, C33	1000-pF polystyrene, 100 V
C40	10-pF variable
C36	15-pF variable
L1, L2	16–30 μH
L3, L4, L11	10–18 μH
L5	20 turns each, wound bifilar on T50-6 core
L6, L7, L10	28-63 μH
L8	100 μH
L9	330 μH

FIGURE 8-5. Continued.

band and 7.0 to 7.25 MHz on the 40-m band. Frequency stability is ensured by using polystyrene and mica capacitors in the frequency-determining circuits.

The local oscillator circuit shown is for a dual-gang 50-pF capacitor. This is not an unusual capacitor and can be found easily from a surplus or used source. However, it is possible to use other values and styles of capacitors for the local oscillator. If a capacitor that is quite a bit different from the type shown is used, it would be desirable to experiment with the oscillator before the circuit is installed in the receiver.

The bulk of the signal processing—IF amplification, detection, and AGC—is handled by an LM373 IF subsystem chip. The AGC action of this chip is so effective that no RF gain control is used except a 20-dB attenuator at the antenna input to increase the dynamic range of the receiver for strong signals.

A single-chip audio amplifier is used, which provides plenty of audio for either loudspeaker or headphone use.

The power requirement is 12 V for the entire receiver, which allows the unit to be operated from storage batteries or from the self-contained 12-V regulated supply.

All the receiver electronics are constructed inside an aluminum chassis 5 by 10 by 3 inches which is fastened to the front panel with 1-inch threaded spacers. Flush-mounting studs are used to fasten the spacers to the front panel without visible hardware. The space between the front panel and the chassis is occupied by controls and the 6-inch-diameter dial plate. The dial assembly consists of a 6:1 planetary reducer mechanism and a 6-inch plastic hand-calibrated dial that were removed from an old signal generator and reworked. The original calibrations were removed from the dial plate by sanding the legends from the plate with fine sandpaper and remarking the dial with an ink pen and dry-transfer letters. The arrangement of the dial-drive mechanism will depend mostly on what is available to the builder. Whatever the mechanical arrangement, the drive mechanism must be smooth and free from backlash. It is a good idea to drill oversize mounting holds for

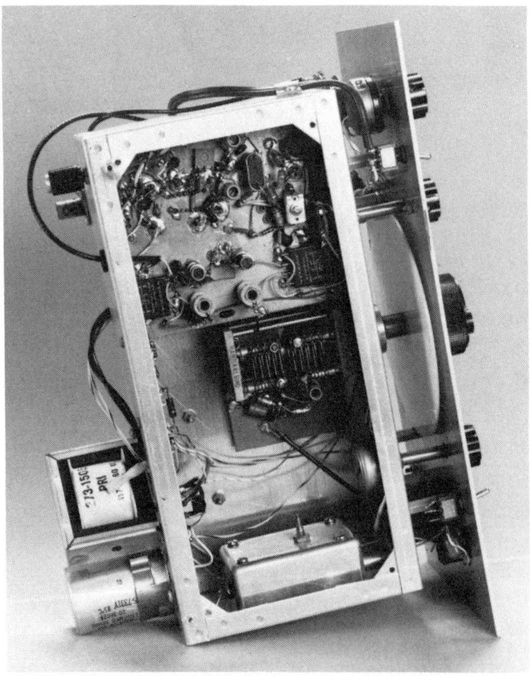

The superheterodyne receiver is an example of the use of a recessed chassis.

the dial components and to align the parts carefully before tightening the hardware.

Inside the chassis, the electronic components are mounted on three copper-clad boards using the techniques described in Chapter 7. A board 3 by 3 inches is used for the local oscillator, and a board 2 by 3 inches is used for the BFO. The remaining electronics, the front-end filters, the mixer and crystal filter, the IF amplifier, and the audio amplifier are constructed on a board 5 by 5 inches. The dimensions of the copper-clad boards used are not critical and the actual size can be determined by the parts size. As always, the lead length should be kept to an absolute minimum to prevent any circuit instabilities.

The BFO is constructed inside a shielded box to prevent harmonics of the BFO from causing spurious signals to be generated by the receiver.

The power supply components are mounted on the rear of the chassis, both on the inside and outside of the chassis.

The IC pin numbers shown on the schematic for the MC1496 are for the round can. The dual-in-line IC will work perfectly well except that the pin numbers are slightly different. Similarly, the pins shown for the LM373 are for the dual-in-line. The round can LM373 is an equivalent chip, but the pin numbers are slightly different. If integrated circuits other than those shown are used, be sure to check a data book for the proper pin arrangement.

Perhaps the most difficult part of receiver construction is getting the completed receiver to work properly. The more test equipment that is available, the easier the job will be. However, it is possible, with care and patience, to troubleshoot and align the receiver with nothing more than a communications receiver and a voltmeter.

Before any alignment is attempted, some preliminary checks should be performed. After the usual checks for wiring errors, apply the power and check for obvious signs of overheating or other damaged parts. If all looks well, check pin 14 of the LM373, pin 14 of the LM380, the drain of Q1, the drain of Q2, and the drain of Q3 for 12 V dc. Measure the gate of Q2 and Q3 for a negative voltage, which indicates that those stages are oscillating. Switch the band switch and check if the relays are operating.

Preliminary dial setting and calibration will be done with a communications receiver. The receiver used need not be calibrated accurately because only preliminary adjustments are being performed. More accurate adjustments will be made with a crystal calibrator. Set the tuning capacitor and C30 for unmeshed position and listen for the local oscillator signal with the communications receiver. A short wire connected to the antenna terminal of the communications receiver should produce a strong signal in the receiver. The correct frequency for the unmeshed capacity of the tuning dial should be 5.50 MHz. If the frequency of the local oscillator is too high, turn the slug of L11 into the coil until the frequency is correct. If the frequency of the local oscillator is too low, turn the slug out of the coil until the frequency is exactly 5.50 MHz. If the frequency of the oscillator cannot be brought down to the proper frequency, increase the value of C29 by about 10 percent or more until the proper unmeshed frequency can be obtained. Similarly, if the unmeshed frequency cannot be brought high enough, decrease the value of C29 by about 10 percent, until the unmeshed frequency can be set. The need to change C29 does not indicate a failure or a mistake in the local oscillator circuit, but is a method of tailoring the oscillator to an available tuning capacitor.

Turn the tuning capacitor to full mesh and locate the signal with the communications receiver. Set C31 so that this frequency is 5.25 MHz.

The frequency-setting steps will be repeated using the crystal calibrator in the receiver after the receiver has been aligned and is operating completely.

Using the communication receiver, listen for the BFO at about 1.75 MHz. It may be necessary to move the antenna wire much closer to the BFO to obtain a strong signal. Many communications receivers do not cover 1.75 MHz; in that case, listen for the second harmonic of the BFO at about 3.5 MHz. With the pitch capacitor set to midrange, adjust L10 for a BFO frequency of 1.75 MHz.

Place the band switch in the 80-m position and switch out the attenuator. Attach a signal generator or an antenna to the receiver antenna terminals. If the antenna is used for alignment, best results will be obtained at night, when the signals are very strong. Tune in the signal generator or an off-the-air signal anywhere in the band and adjust L6, L7, and C10 for maximum signal level. Since the S-meter will not be operating properly at this time, the adjustments will be by ear. Select a signal at about midrotation of the dial and adjust L1 and L2 for maximum signal strength. Slowly tune the receiver through a strong signal anywhere in the band and adjust C14 for best selectivity.

Change the band switch to 40 m and tune in the signal generator or off-the-air signal near 7.125 MHz and tune L3 and L4 for maximum signal strength.

The local oscillator will be adjusted and the dial calibrated in the next steps using the internal crystal oscillator. Set the tuning capacitor to slightly less than full mesh. Place a pencil mark on the dial to mark the location and tune C31 to zero beat, the calibrator signal. Turn the capacitor to tune in the crystal marker at 7.100 MHz and place a pencil mark at the point of zero beat. Turn the tuning control to the third crystal marker at 7.200 MHz. Mark this point with a pencil. If the oscillator has been adjusted correctly, the 200 kHz of the dial should occupy about 150 degrees. If the bandspread is adequate, the dial can be inked, the dry-transfer letters can be applied, and the dial can receive a coating of clear protective lacquer. Since the same local oscillator frequency range is used for both the 40- and 80-m bands, the calibrations obtained for the 40-m band ensures accurate calibration on the 80-m band.

With the antenna removed, set RB so that the S-meter is about midscale. Adjust RA so that the S-meter just begins to move to the right. Readjust RB so that the S-meter reads zero.

This completes the alignment of the superheterodyne receiver. If the receiver is used with a transmitter, a transmit/receive switch connection can be made to the terminals marked "T/R switch." The connection should be such that the contacts are opened during periods of transmit and closed during periods of receive.

8.5 TWO-BAND SOLID-STATE TRANSMITTER

This project describes the construction of a two-band VFO-controlled CW transmitter (Figure 8–6) designed and constructed as a matching unit for the two-band superheterodyne receiver described earlier. This transmitter uses the frequency-division method of band switching. That is, the VFO operates at a frequency above the highest frequency the transmitter operates and the frequency is divided by integers with digital logic to reach the desired amateur band. In this case, the transmitter covers the 40-m band as well as the 80-m band. The VFO operates from 14.0 MHz to about 14.4 MHz and is divided by 2 to generate the 40-m CW band coverage of 7.0 to 7.2 MHz. This is divided by 2 again to obtain the lower portion of the 80-m CW band of 3.5 MHz to 3.6 MHz. The remainder of the 80-m band is generated by switching a capacitor out of the VRO circuit, raising the frequency of the VFO of 14.4 MHz to 14.8 MHz. After dividing by four, this frequency becomes 3.6 to 3.7 MHz, which is the upper portion of the 80-m CW band.

TTL integrated-circuit logic is used to generate the required signals as well as to switch between bands. An important property of digital flip-flops, which are used to generate the frequency division, is used to provide the proper phase relationships for the push-pull amplifiers. All logic flip-flops produce two output signals. One signal is called Q and the other signal is called $\bar{Q}$, which is pronounced Q-not. In radio-frequency language, this means that the Q and $\bar{Q}$ signals are out of phase, which is required for push-pull amplifiers. Using this relationship, the Q and $\bar{Q}$ signals are switched through a digital multiplexer and

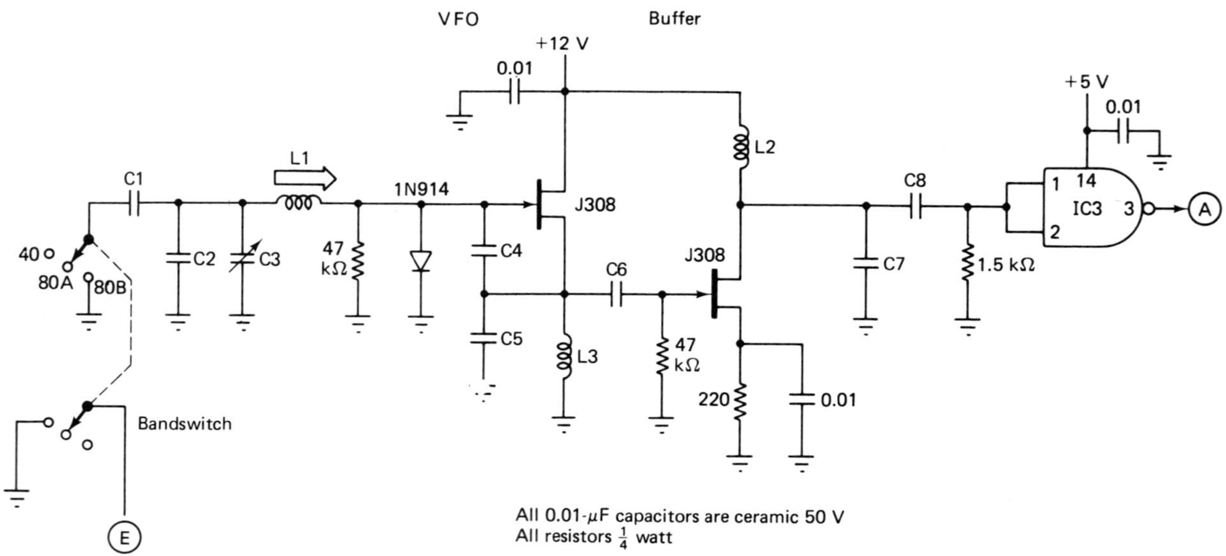

FIGURE 8–6. Schematic of the two-band transmitter.

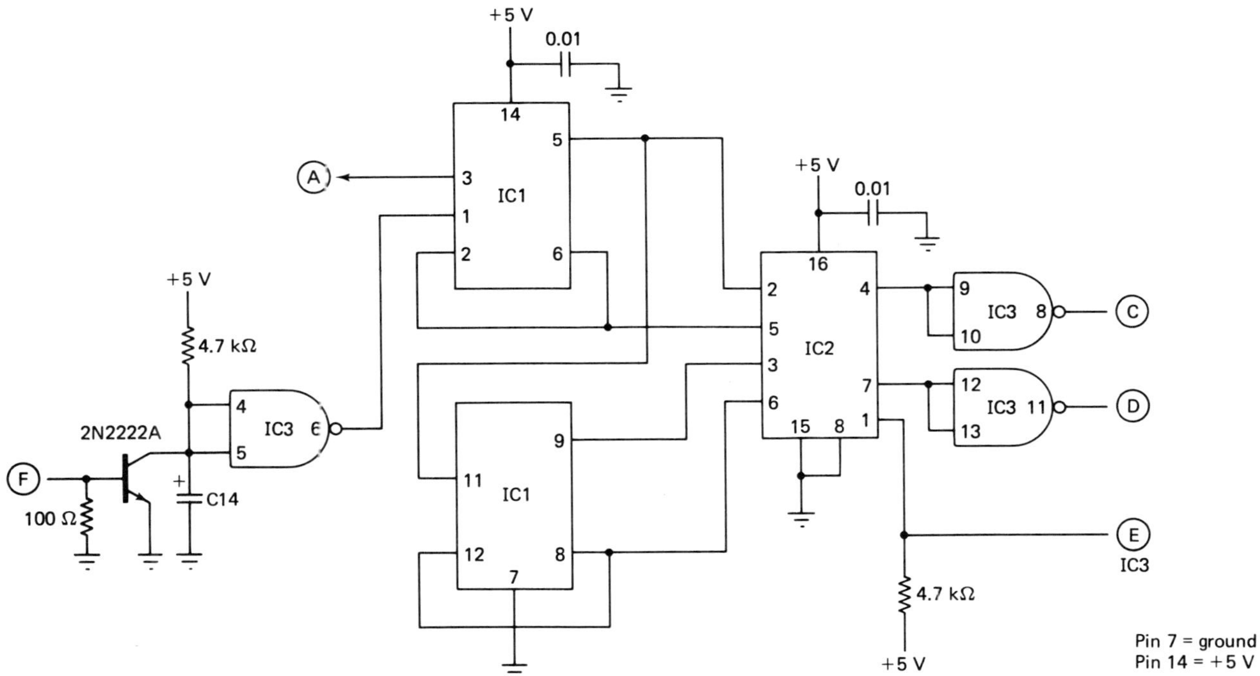

FIGURE 8-6. Continued.

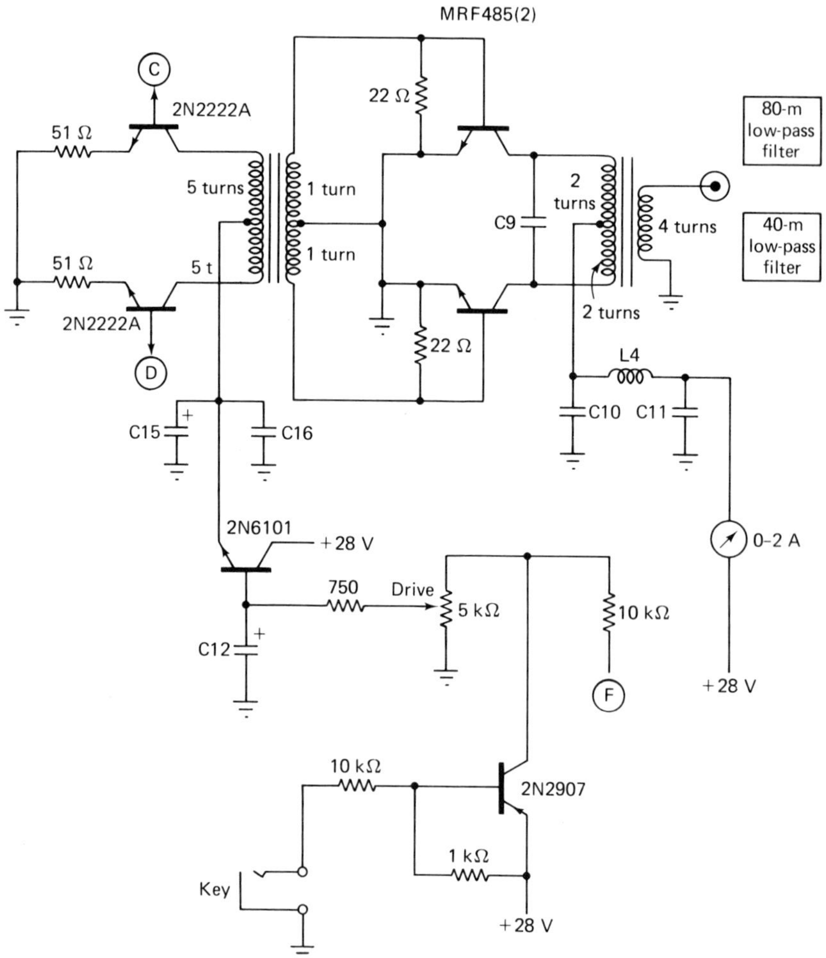

FIGURE 8-6. Continued.

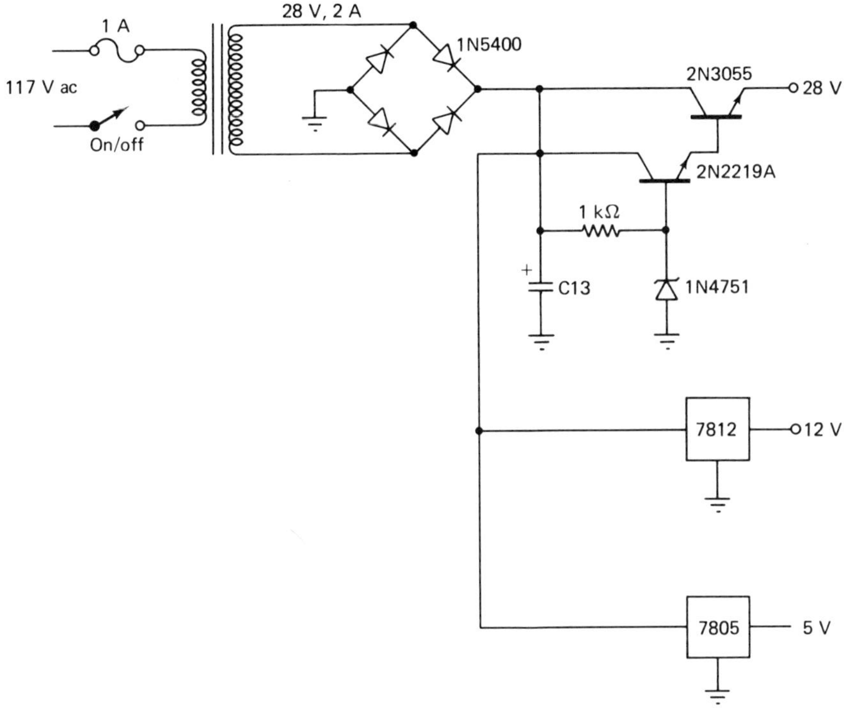

Two-Band Transmitter

C1	100-pF compression trimmer
C2	100-pF mica, 50 V
C3	50-pF variable capacitor
C4, C5	470-pF mica 50 V
C6, C7	43-pF mica
C8	4.7-pF mica
C9	250-pF ceramic
C10, C16	0.01-μF ceramic
C11, C15	1-μF tantalum, 35 V
C12	2.2-μF tantalum, 35 V
C13	22, 000-μF computer-grade electrolytic, 50 V
C14	100-μF tantalum 35 V
IC1	7474
IC2	74LS157
IC3	7437
L1	1.5–2.0 μH
L2	4.7-μH RF choke
L3	330-μH RF choke
L4	Two ferrite-shield beads

FIGURE 8–6. Continued.

A good match for the superheterodyne, the two-band CW transmitter is also housed in a wooden cabinet.

are used to drive a two-stage push-pull amplifier. There are no tuned circuits in the push-pull amplifier stages, so that no switching is required between bands, but a low-pass filter (Figure 8–7) is provided to remove the harmonic distortion produced in the amplifier.

A time-sequenced keying system is provided to provide a smooth, click-free keying shape. The keying is shaped through the modulation of the first push-pull stage. The collector supply voltage is modulated with a controlled rise time which shapes the drive to the power-output stage. The keyer also assures that the digital logic has been enabled before the modulation is applied to the driver and that the logic continues to operate after the keying envelope has reached zero. If the digital logic were allowed to turn off too soon, there would be a very abrupt cessation of the RF waveform and a resultant serious key click. The collector supply voltage is applied to the power amplifier at all times, which operates as a class C amplifier.

The power supply for the transmitter contains a number of voltage regulators. First, the 28-V collector supply for the power amplifier and the driver is regulated. It would not seem necessary to regulate the power supply for a high-power RF amplifier. However, to operate a transistor amplifier at, or near, its maximum rated power with safety it is necessary to provide a regulated supply voltage. If the collector

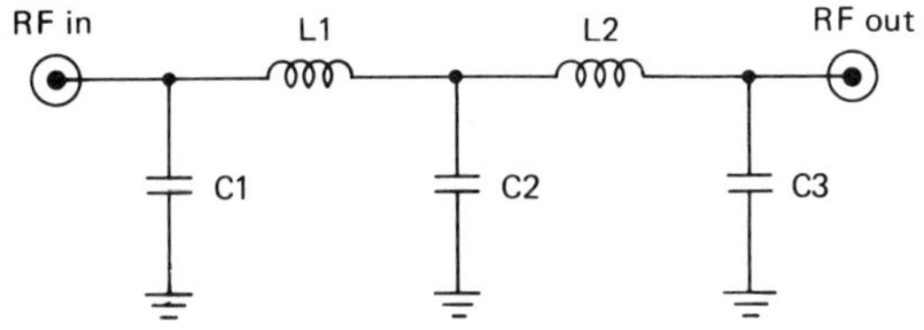

80-Meter		40-Meter	
L1, L2	24-turn T50-10	L1, L2	19-turn T50-6
C1, C3	1000 pF	C1, C3	500 pF
C2	1800 pF	C2	820 pF

T50-10 and T50-6 toroid cores are manufactured by Micrometals and other manufacturers under the same number.

FIGURE 8-7. Low-pass filter for the VFO transmitter. These filters should be built in an aluminum box.

supply were from an unregulated supply, the voltage would droop during transmit, which would not allow for full power output. To achieve full power output, the power supply droop must be accounted for by raising the key-up power supply to a slightly higher voltage, which means that the power supply voltage will be above the rated voltage before the key is depressed. When the key is first depressed and before the power supply has the time to droop, the power amplifier is operating beyond its specifications. Although the time is short, it may be long enough to destroy the transistors. Therefore, a simple regulator is used for the 28-V supply line. A regulated 12 V is supplied for the VFO while another regulator for the 5-V logic is also supplied. The dial lamp and the antenna change over relay are surplus 28-V aircraft units and operate from the regulated 28-V supply.

The transmitter is constructed on a standard radio chassis attached to an aluminum panel by the use of threaded spacers and flush-mounted blind fasteners. The chassis is set back from the front panel about 1 inch so that a large-diameter dial can be placed directly behind the panel. The dial is hand-calibrated using dry-transfer letters as described in Chapter 7. The digital-logic integrated circuits are all contained on a 2 by 3 inch piece of universal DIP board and mounted on the rear apron of the chassis close to the power amplifier, which is mounted on a heat sink on the outside rear apron of the chassis. It is important to keep the leads within the wiring of the power amplifier extremely short to prevent parasitic oscillations or impedance mismatches. The only parts of the transmitter that are not mounted on the chassis are the power supply and the low-pass filters. The separate housings for the low-pass filters does not mean that they are in any way an optional part of the transmitter design. The spectral purity of the transmitter output is not clean enough to be used directly on the air without a low-pass

A close-up of the power amplifier showing the broad-band transformers.

filter, even if an antenna tuner is used. The low-pass filters are built into two separate boxes with connectors and are inserted between the transmitter and each antenna. This puts one low-pass filter on the 80-m antenna and the other on the 40-m antenna. Switching antennas accomplishes the task of switching the low-pass filter as well. In addition, the low-pass filter is also inserted in the receiver input for additional protection from receiver intermodulation. It would be permissible to mount the low-pass filters on the transmitter chassis and switch them with the band switch. Regardless of how the filters are implemented, they must be used with the transmitter.

Because of the small number of tuned circuits in the transmitter, the unit is very easily aligned. Before any calibration of the VFO may be performed, it must be determined that the transmitter is operating correctly. Incorrect operation can cause very heavy currents to be drawn from the power supply and can upset regulation of the power supply. After the transmitter has been constructed and checked for wiring errors, connect the key and a dummy load to the transmitter. If the low-pass filters were constructed in individual boxes, they need not be used for these off-the-air tests. Apply the ac power and check the three regulated power supply voltages. The power amplifier collector voltage should be 28 V plus or minus 3 V, the VFO supply should be 12 V plus or minus 2 V, and the logic voltage should be 5 V plus or minus 0.5 V.

The VFO dial will be hand-calibrated using the techniques outlined in Chapter 7, which requires a communication receiver with a well-calibrated dial or a crystal calibrator. The VFO operates in the range 14 to 14.8 MHz and the communications receiver must cover this frequency range. Since the VFO operates continuously, it can be heard in the communication receiver without pressing the key. It may be necessary to place a wire from the communications receiver antenna jack to the vicinity of the VFO to obtain a signal strong enough to be heard in the receiver.

Place the band switch in the 80-m B position and turn the VFO tuning control until the VFO variable capacitor is almost at full mesh. Mark the dial at this point and turn the slug in L1 until the VFO frequency is exactly 14.4 MHz. Rotate the VFO tuning control and place pencil marks at 14.5, 14.6, 14.7, and 14.8 MHz. These points will be marked as 3.5, 3.625, 3.65, 3.675, and 3.700. The capacitor should be

The low-pass filters require a shield between the input and output to prevent higher frequency harmonics from leaking through.

about 80 percent unmeshed at 14.8 MHz. If it is not, a fixed capacitor of about 20 percent of the value of the variable capacitor should be placed in parallel and the procedure repeated.

The band switch should be placed in the 80-m A position and the capacitor set in the same position as before with the plates almost meshed. C1 is turned until the VFO frequency is exactly 14.0 MHz. The VFO dial is turned and pencil marks are placed at exactly 14.1, 14.2, 14.3 and 14.4 MHz. These marks correspond to 3.5 and 7.0 MHz, 3.525 and 7.05 MHz, 3.55 and 7.1 MHz, 3.575 and 7.15 MHz, and 3.60 and 7.2 MHz. If the calibration was performed correctly for the 80-m B position, it should be possible to obtain this range for 80-m A position. The dial is completed and the front panel is assembled.

Pressing the key should produce a current reading on the power amplifier current meter. Full drive provides about 1.5 to 2 A of collector current. Low collector current indicates insufficient drive or a mismatch in the antenna circuit, whereas high current usually indicates a mismatched circuit in the antenna. Usually, the current obtained on the 40-m band is less than the current obtained on the 80-m band. The correct currents for this transmitter are about 1.75 A on the 80-m band and about 1.25 A on the 40-m band. This corresponds to about 25 W output on 80 m and about 17 W output on 40 m.

9 CONSTRUCTION PROJECTS: ACCESSORIES

9.1 THREE-FREQUENCY CRYSTAL CALIBRATOR

A crystal calibrator was usually built into older high-quality receivers to assure their dial accuracy. Just about the only frequency used was 100 kHz, which implied that the dial could be set exactly on frequency at multiples of 100 kHz. Between the calibration points, the accuracy of the dial was a function of the precision of the dial mechanism. Even in good communications receivers the error in the midpoint of the 100 kHz spread could be substantial. Having a calibration point closer than 100 kHz could reduce this error substantially, and thus 50 kHz and even 25 kHz calibrators are now available. This project is a three-frequency (100, 50, and 25 kHz) crystal-controlled calibrator.

The heart of the crystal calibrator (Figure 9–1) is a CMOS integrated circuit that is usually used for generating a time base from a crystal oscillator. The integrated circuit has an uncommitted transistor that may be wired as a Pierce crystal oscillator. The 3.2-MHz crystal is divided by a factor of 32, which is five binary flip-flop frequency dividers, to produce the 100-kHz output. The 3.2-MHz crystal has two important advantages over the usual 100-kHz unit. The 3.2-MHz crystal is smaller and considerably less expensive, with no loss in performance. The outputs from the two next stages are also used and are 50 and 25

Crystal calibrators in a small box. This unit has a self-contained battery and has three markers.

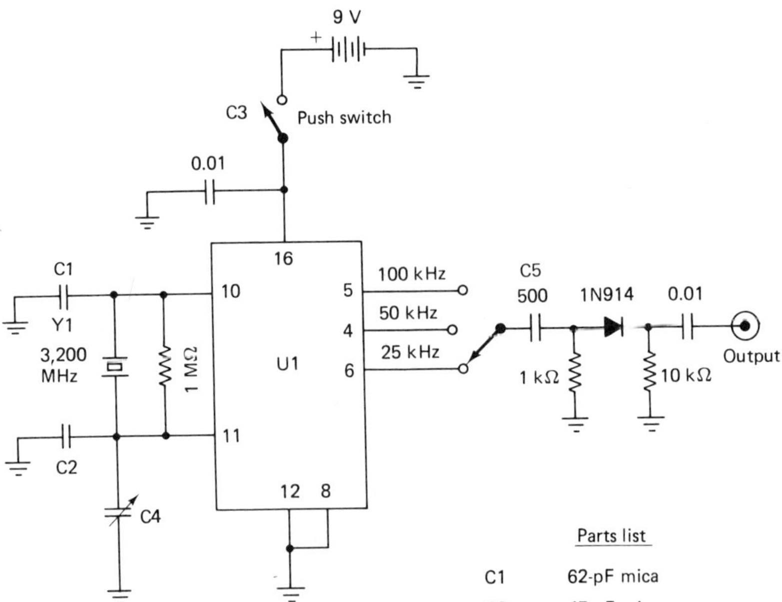

Parts list

C1	62-pF mica
C2	47-pF mica
C3	0.01-μF ceramic
C4	7-35-pF ceramic trimmer
C5	500-pF mica

FIGURE 9-1. Schematic of the crystal calibrator.

KHz. Since the outputs are square waves, there should be no even harmonics. This would make it difficult, if not impossible, to hear the 36th harmonic at 3.6 MHz in the 80-m band. On the other hand, the 37th harmonic would be strong at 3.7 MHz. This weakness is removed by changing the waveshape to a spike with a fast signal diode and capacitive coupling to the output. This wave shape has harmonics to a very high frequency, with a constant signal strength for each harmonic. Because of the desire to have harmonics up to and including at least the 300th at 30 MHz, the CMOS must be operated from at least 9 V. Many CMOS chips will operate at 15 or even 18 V, and this would produce even more improved operation. A small transistor radio battery will provide long operation life, since the calibrator is used for only short periods of time in addition to the low battery drain of the CMOS circuits. A pushbutton switch is used rather than a toggle or rotary switch, to prevent the unit from inadvertently being left on and depleting the battery.

Because of its simplicity, the crystal calibrator was constructed on a small piece of perforated stock. A PC board could be made for the calibrator, but for a circuit this simple, it hardly seems worth the effort.

The front panel contains a rotary switch to select the marker output frequency, 25, 50, or 100 kHz, while the rear apron contains only a jack for the signal output. No power supply or power cord is required, as the unit is operated from self-contained batteries. The overall dimensions of the unit are $3\frac{1}{4}$ by $2\frac{1}{8}$ by $1\frac{1}{2}$ inches.

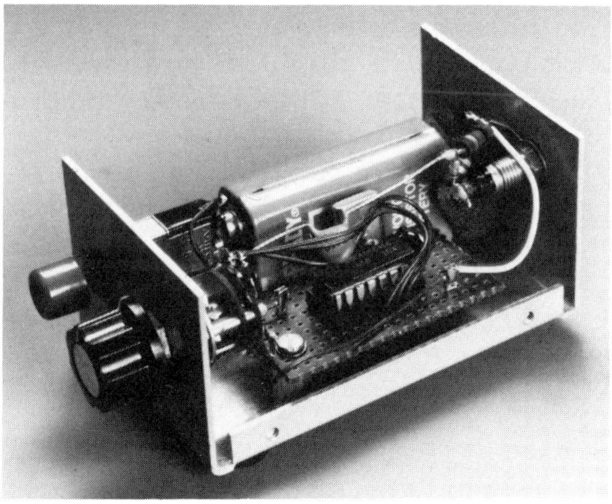

A small piece of perforated board easily holds all of the circuits required.

After the calibrator is constructed, it should be checked for operation. Set the frequency switch to 100 kHz and press the button while using a receiver to listen for one of the harmonics of the calibrator. If a communications receiver is used, sufficient signal can be obtained by using a short piece of wire in the receiver antenna jack, positioned about 1 foot from the calibrator box. Many of the lower-order harmonics should be heard and certainly the crystal oscillator at 3.2 MHz should be heard. Another good tool for determining proper operatin is a portable AM broadcast-band radio. A wire from the calibrator output can be positioned near the rod antenna of the broadcast receiver and the calibrator activated. Once it has been established that the calibrator is operating, the frequency must be trimmed to the correct value. This is done by beating the 10-MHz harmonic to WWV transmissions. To accomplish this, the calibrator must be connected to the receiver in the normal fashion. Sufficient energy should be available from wrapping about 3 inches of the lead from the calibrator around the antenna input lead in the receiver. If the receiver has a shielded antenna connector, a tee fitting may be used and the calibrator coupled in with a 1- to 5-pF capacitator. If the receiver is a part of a transceiver, sufficient signal strength may be obtained by placing the wire from the calibrator around the RF amplifier tube.

Once the calibrator is connected, tune in WWV or any of the standard frequency stations operating on frequencies that are direct multiples of 100 kHz. Place the calibrator switch to 100 kHz and press the calibrator button. A beat note should be heard between the calibrator and the standard frequency station. Turning C4 should cause the pitch of this beat note to change. Turn the capacitor slowly to reduce the frequency of the beat note to as low as possible. As the beat approaches very low frequencies, it may not be possible to hear the beat. In some cases the beat frequency can be seen on the S-meter, and this may be used as an indication of zero beat.

It is important to tune the calibrator as close to the actual frequency as possible. If the beat note is 50 Hz at 5.00 MHz, the error at 25 MHz will be 250 Hz. The greatest accuracy will be obtained if the calibrator is zero beat to a high-frequency WWV transmission. Only the 100-kHz marker need be calibrated because the 50- and 25-kHz markers are derived from the same source.

9.2 100-WATT LINEAR AMPLIFIER

The two transmitters described in Chapter 8 provide between 8 and 25 W of output, which can provide many contacts and provide a real challenge for DX work. For more serious DX work or just winning the fight

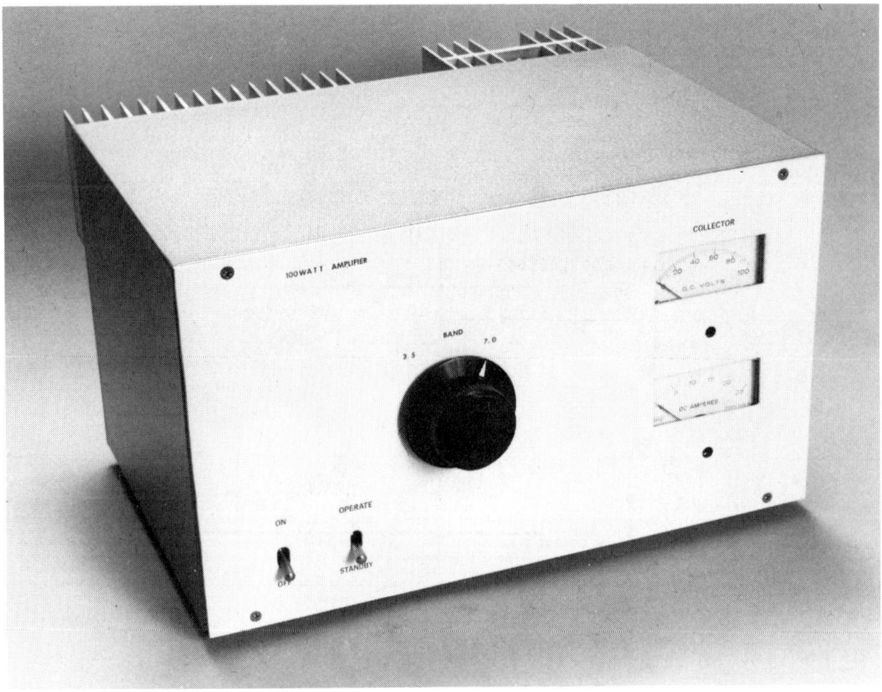

The power amplifier covers two bands to match the transmitters described in Chapter 8.

against QRM, a higher power level is required. This amplifier can provide more than 100 W of output on the 80- and 40-m bands using two transistors. Less then 4 W of drive is required for full output power. This low power level would cause the VFO transmitter described in Chapter 8 to overdrive the amplifier and the two-band transmitter's higher output would potentially damage the unit. To bring the drive requirement of the amplifier more in line with the power output of the two transmitters described in Chapter 8, a 3-dB attenuator is included in the amplifier circuit. This produces an amplifier that has an 8-W drive requirement for full power output.

The amplifier (Figure 9–2) is a single-stage transformer-coupled push-pull circuit using two 50-W transistors. RF feedback is employed to stabilize the circuit gain to prevent oscillations and to present a more constant drive requirement. The 3-dB attenuator also helps stabilize the input impedance to 50 Ω. The input and output transformers are constructed on ferrite cores using copper tubes and printed-circuit-board material as described in Chapter 8. The input attenuator is constructed using low-inductance carbon resistors arranged as three 1-dB attenu-

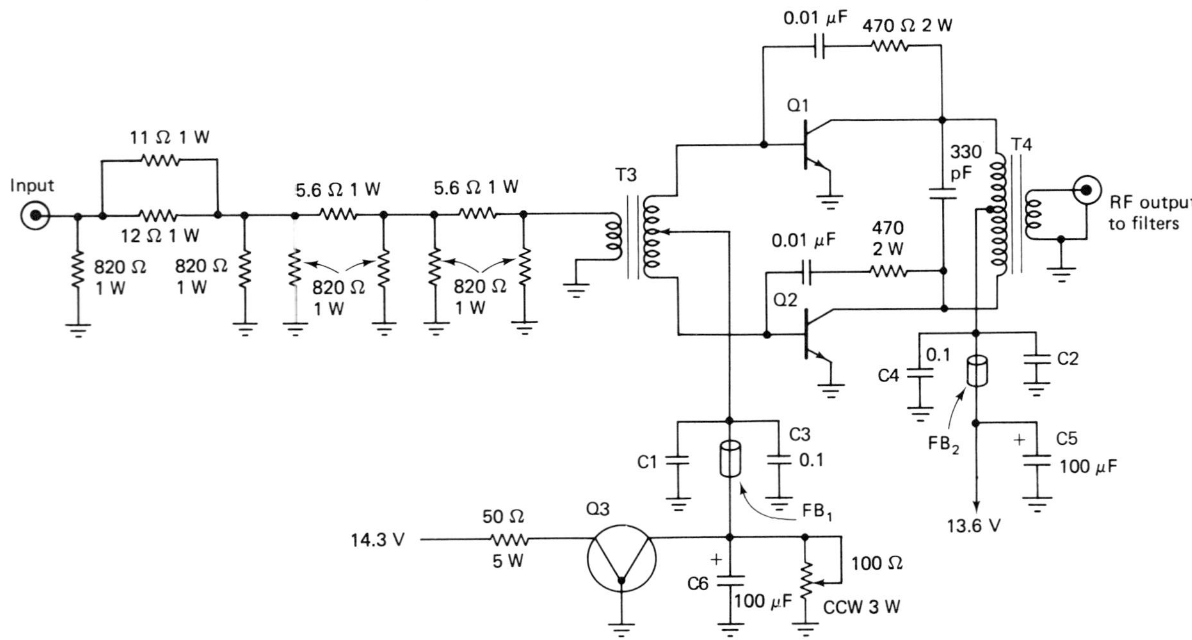

FIGURE 9–2. Schematic of the 100-W amplifier.

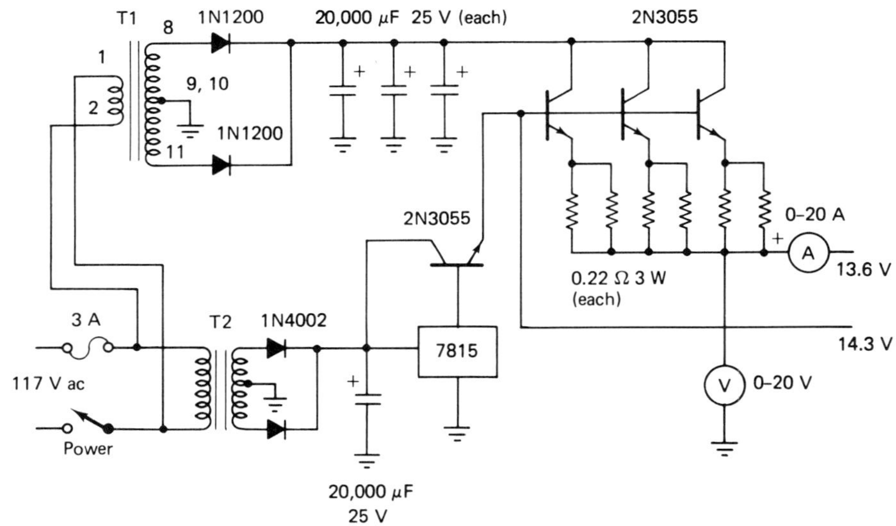

FIGURE 9–2. Continued.

100-Watt Amplifier	
C1, C2	560-pF Underwood Electric Corp. uncased mica
C3, C4	0.1-μF ceramic, 50 V
C5	100-μF tantalum electrolytic
C6	1000-μF tantalum electrolytic
C7, C9	470-pF mica, 300 V
C8	820-pF mica, 300 V
C10, C12	910-pF mica, 300 V
C11	1500-pF mica, 300 V
FB1	Ferrite-shield bead
FB2	Ferrite-shield bead
L1	15 turns on T80-6 Micrometals core
L2	15 turns on T80-6 Micrometals core
L3	24 turns on T80-6 Micrometals core
L4	14 turns on T80-6 Micrometals core
T1	RT 208, Stancor
T2	28 V center-tapped, 1 A
Q1, Q2	S-50-12 Transistor, Communications Transistor Corp.
Q3	BY-1 "Byistor," Communications Transistor Corp.
T3	Wide-band transformer made with four beads as described in text; one turn secondary, four turns primary
T4	Wide-band transformer made with 10 beads as described in text; one turn primary, four turns secondary.

FIGURE 9-2. Continued.

ators in cascade. Do not, under any circumstances, substitute wire-wound resistors for any of the carbon units. The inductance of the wire-wound resistors can cause damage to the amplifier, the driver, or both, from an excessive standing-wave ratio.

The output transistors are biased to about 30 mA each using a temperature-compensating voltage source mounted directly on the amplifier heat sink. This technique was discussed in Chapter 2, which described its ability to prevent thermal runaway. The device used for generating the bias voltage source is a special resistor/diode combination made specifically for solid-state power amplifiers.

A low-pass filter (Figure 9-3) is used for each band and selected with the front-panel band switch. The filters are five-pole filters using toroid inductors and mica capacitors. The filters are required to lower the harmonic energy to a safe and legal level. The filters are placed in a shielded compartment to decrease the likelihood of energy escaping and being radiated directly from the filter.

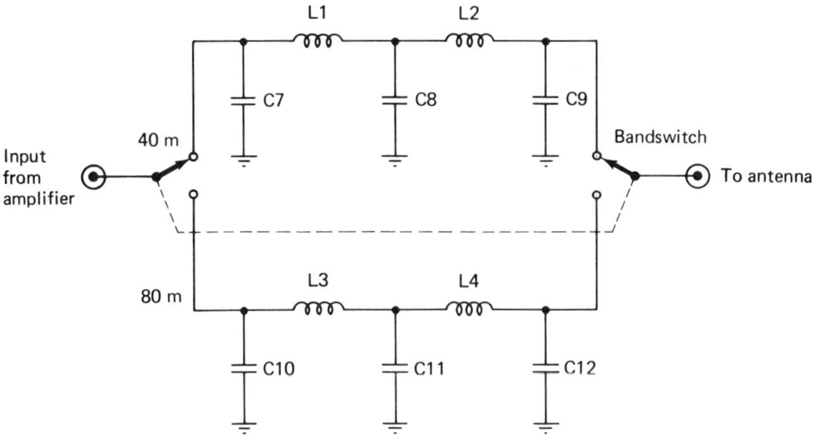

FIGURE 9-3. Contents of the filter box.

The power supply provides 13.6 V regulated at 15 A, which is capable of providing the amplifier with a sustained 200 W of input, which will produce over 100 W of power output. The voltage regulation is not as stiff as a full closed-loop feedback system. The regulation is provided to allow the transistors to be operated at their maximum supply voltage without danger of damage when the supply voltage rises. The regulator prevents the supply from exceeding 13.6 V during periods of standby. Three inexpensive 2N3055 transistors are wired in parallel with current-sharing emitter resistors mounted to a large heat sink on the rear of the amplifier cabinet. These three are driven from another 2N3055, which is ultimately driven from a three terminal voltage regulator. A separate power supply with a higher voltage than the main supply is used to supply the base-bias circuits, the three-terminal regulator, and the 2N3055 transistor, which drives the three transistors in parallel. This allows the voltage differential across the 15-A regulator to be at a minimum for a safe temperature rise.

The amplifier cabinet houses three basic systems: the amplifer itself, which is mounted on a heat sink mounted on the back panel; the power supply, which is distributed throughout the cabinet; and the low-pass filters, which are mounted in an aluminum box mounted behind the bandswitch.

When constructing the unit, great care must be taken when constructing the power amplifier. The parts layout must be exactly as shown in Figure 9-4 if the amplifier is to be constructed without instabilities. The amplifier is constructed on double-sided printed-circuit-board material. The side toward the heat sink is solid copper and contains no components. Copper straps are soldered to both sides

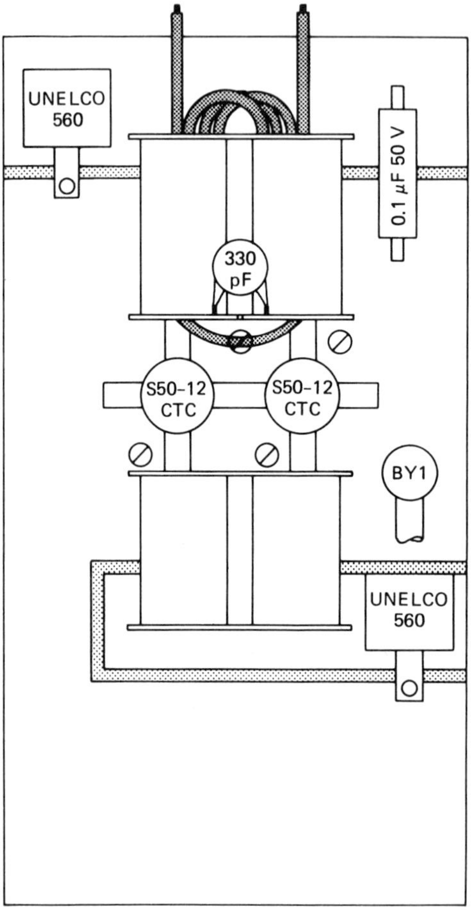

FIGURE 9-4. Layout of the 100-W amplifier.

of the board to ensure that both sides of the board are connected together. In addition, holes are drilled in several spots and wires are soldered to the copper foil on both sides of the board. The pattern used for the amplifier is so simple that a printed circuit board could be created by simply cutting the copper with a sharp knife and peeling away the unwanted foil. Many of the components shown are special low-inductance components and should not be substituted for by other components, even though the values are the same.

When wiring the power supply, be sure to use wire of sufficient size to pass the desired current. Number 12 or 14 wire should be used for those leads that are to carry 15 A. Do not rely on chassis grounds to

Rear view of the amplifier showing the four transistors of the power supply.

pass the ground currents. Both the positive supply and the negative return lead should be copper wire.

After the amplifier has been constructed, it should be checked for wiring accuracy and tested. Whenever working with the amplifier, a suitable dummy load should be attached to the output. Although the amplifier should be capable of operating into an infinite VSWR, it is not desirable to work on the amplifier under this condition. Before applying power, set the bias control full counterclockwise, or minimum bias. A variable transformer would be useful for the initial check, as it allows the line voltage to be applied slowly. Without drive, and without bias, apply ac power to the amplifier. If the voltmeter does not read about 13.6 V and the ammeter reads zero, remove the power immediately and check for wiring errors or damaged components. If the proper meter readings have been obtained, advance the bias control while watching the ammeter. The current should rise smoothly without any sudden jumps or excursions. Do not exceed 1 A during this test. If the bias control is smooth and there are no indications of instabilities, set the bias current to 60 mA. It may be difficult to read 60 mA accurately on the 20- or 25-A meter used on the amplifier. In this case, an external milliampere meter may be used to make this adjustment.

Once the quiescent current has been set and the power supply voltages are correct, apply the drive to the amplifier. Be sure that the bandswitch on the power amplifier is set to the same band as the driver. If the VFO transmitter from Chapter 8 is being used for the test, no control of the drive level is possible and the full drive will be applied. If the two-band transmitter is being used to drive the amplifier, the drive-level control on the transmitter should be started at full counterclockwise and advanced slowly while the key is depressed. In no case should there be any indications of unstable currents or power supply voltages. The power supply voltage will sag a little when full drive is applied. When operating at full drive, the collector voltage will be between 12 and 13 V and the collector current should be between 15 and 16 A. Only enough drive to achieve this current should be applied. Excessive drive only increases the possibility of damage and increases the levels of the spurious outputs.

Connect a wattmeter to the amplifier output and measure the power output. With 200 W of input, the output power should be in excess of 100 W. If this power is not available, check the SWR of the dummy load. If the power output is less than 100 W, do not operate the amplifier. Low efficiency indicates that power is being dissipated within the amplifier and could cause damage to some of the components.

Certain precautions must be taken to obtain maximum service from the power amplifier. First, any antenna used with the amplifier should be checked for SWR. If the SWR is not less than 2:1, an antenna tuner should be used to tune the antenna to 50 Ω. Increasing the drive to the amplifier can increase the power output beyond 100 W but is not desirable for several reasons. First, the transistors are characterized for 100 output and can suffer damage from excessive drive. Second, the power supply is not designed to supply more than 200 W of input power and will overheat, and finally, the harmonics will increase dramatically and possibly cause excessive heating in the low-pass filters. Since the amplifier is a linear amplifier, single-sideband signals may be amplified with the unit. The maximum power output is 100 W PEP, corresponding to an input drive of about 8 W PEP. If SSB operation is expected, it is important to set the bias current accurately.

The amplifier can be modified to operate other amateur bands by simply changing the low-pass filters. The transistors are capable of supplying 100 W output up to 30 MHz, with a reduction in gain at the higher amateur bands. Values for low-pass filters may be found in the *Radio Amateur's Handbook*, published by the American Radio Relay League. Any high-frequency band may be added by simply adding a low-pass filter with a cutoff frequency slightly above the top end of the desired amateur band. When operating other bands, where the gain may not be as high as on 40 or 80 m, be sure not to exceed the 200 W input power.

9.3 IAMBIC CMOS KEYER

Most CW operators consider the ultimate in keyers those that have an iambic-squeeze-key operation (Figure 9–5). Normal electronic keyers will produce a string of dots when the paddle is pressed to the right and a continuous string of dashes when the paddle is pressed to the left. When using a two-paddle squeeze key, pressing both the left and right half of the key produces a string of alternating dots and dashes. This allows one side to be depressed and without releasing that side, the opposite paddle may be depressed to insert one dot or dash. This reduces operator motions and reduces fatigue. For example, consider the letter Y as sent on a normal electronic keyer. First the keyer is pushed to the left for the first dash, then reversed to the right for the dot and finally back to the left for the remaining two dashes. To send this with the iambic squeeze keyer, the right paddle is depressed for the first dash and not released while the left paddle is also depressed and released. The right paddle remains depressed to the completion of the character.

To take advantage of iambic operation, a two-paddle actuator must be used. Using a normal single-paddle key will produce normal keyer characteristics without degradation.

This keyer features several characteristics that make it a good keyer for serious CW operators. First, the startup delay, that is, the time required for the actual keying to start measured from the time the paddle is actuated, is less than one-eighth of the time of a dot, which is virtually unnoticeable. Dot and dash memories are included to smooth

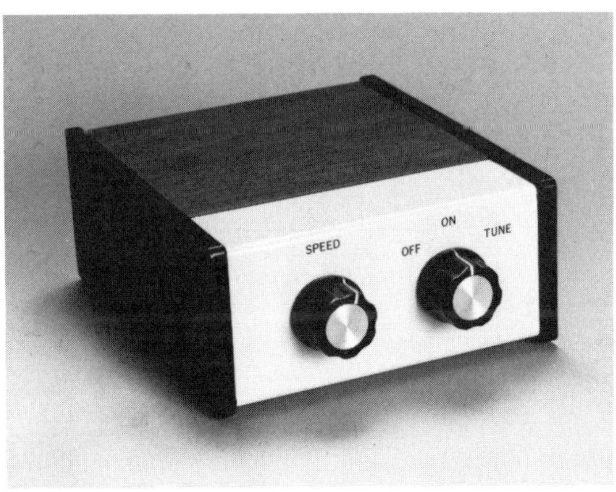

An iambic keyer using CMOS integrated circuits has a very low power drain and could be operated from batteries.

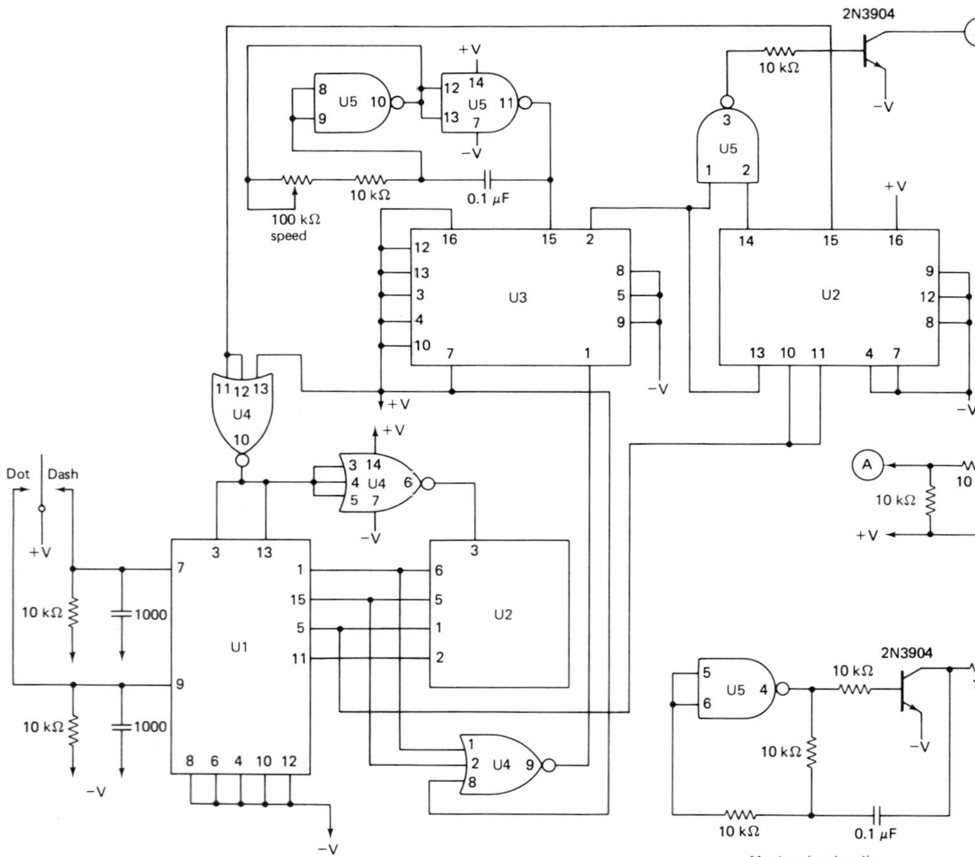

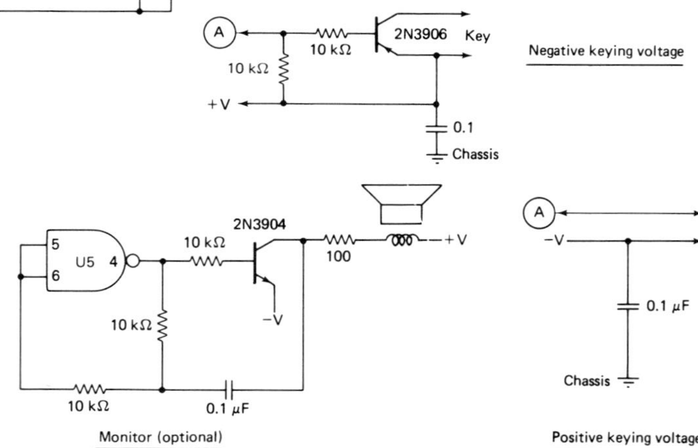

Iambic Keyer

U1, U2	CD4027
U3	CD4516
U4	CD4025
U5	CD4011
R1, R2, R3, R4, R5, R6, R7	10 kΩ, 1/4 W
R8	100 Ω, 1/4 W
R9	100-kΩ linear taper potentiometer
C1, C2, C3	0.001-μF ceramic, 50 V
C4, C5	0.1-μF mylar, 50 V

FIGURE 9-5. Schematic of the iambic keyer.

Interior view showing the five integrated circuits of the keyer.

out the sending of the operator, which tends to get ahead of the keyer. Solid-state keying is included for keying any type of transmitter without the use of relays. The CMOS circuits ensure a long battery life for keyers that rely on battery power for portable work. Also, a monitor may be added if desired.

In spite of all these features, the keyer contains only five integrated circuits and two transistors. The prototype keyer was constructed on a perforated board and point-to-point wiring. Almost any other type of construction could be used because the frequencies involved are low. It would be a good idea to mount the keyer in a metal box and bypass all the power leads if an external power supply is used. Other external connections, such as bypassing for the paddle leads, are shown in the schematic diagram.

If the keyer was wired correctly and there are no defective parts, it should function from the start without any problems. Nevertheless, problems may exist, and it would be a good idea to do the initial testing using a 5-V power supply. CMOS integrated circuits are almost impossible to destroy with a supply voltage as low as this, and the keyer will work perfectly.

9.4 TEN- AND 15-METER PRESELECTOR

One of the improvements that can be applied to an older communications receiver without internal modifications is the addition of a preamplifier for improved sensitivity and image rejection. Many older and

The preselector can be used for the 15- and 10-m bands as well as all frequencies in between.

inexpensive receivers had insufficient RF gain and selectivity on the higher-frequency bands. The addition of the preselector can bring up the performance of the receiver on the higher-frequency bands to rival practically any other receiver available. The sensitivity of an inexpensive receiver on the 80-, 40-, and 20-m bands was usually limited by atmospheric conditions and would not benefit from the addition of a preamplifier.

The preselector (Figure 9-6) consists of a two-FET cascode amplifier using two junction FETs and a two-section tuned circuit. Both the input and the output of the amplifier are high-*Q* circuits tuned with a two-gang capacitor. The frequency coverage of the preselector is contınous from about 18 MHz to about 30 MHz, which covers the 15- and 10-m bands, the new 18- and 24-MHz bands, as well as all other frequencies within that range.

The key to constructing a smooth-working preselector is the tuning capacitor. The capacitor is a two-gang broadcast band variable with identical sections. Many broadcast variables have different values of capacitance for the oscillator and the antenna sections. Capacitors with dissimilar sections may be used by removing plates from the larger section until both halves have the same value, but it may be difficult to achieve tracking. The gain of the amplifier circuit is so much greater than is necessary that any small errors in tracking will not cause any severe degradation in performance. The capacitor used in the preselector shown in the photograph uses an AM/FM radio variable, with the AM and FM sections tied together. Any variable capacitor with identical sections of about 75 to 300 pF may be used.

All the electronic circuitry is mounted on a copper-clad board that is soldered to the variable capacitor frame. This allows for short leads

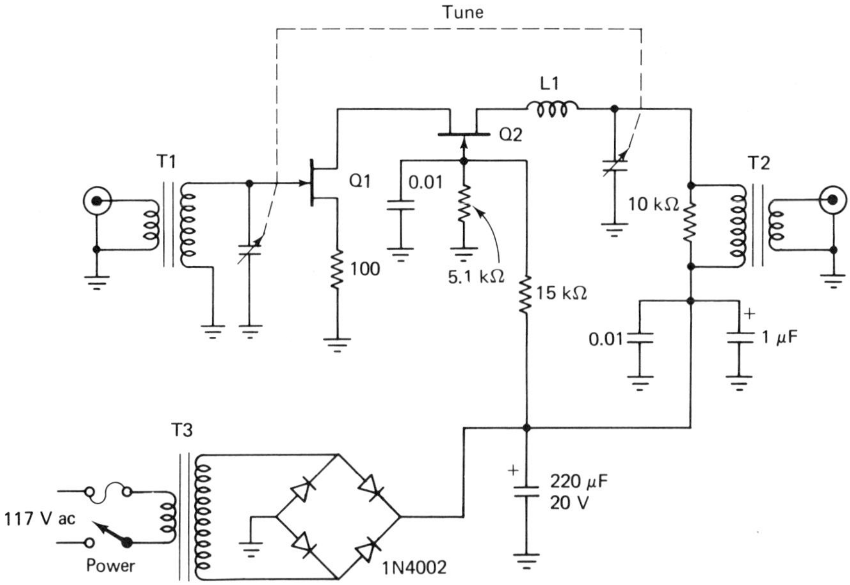

Parts List for 15- and 10-Meter Preselector

Q1	J308, U308, MPF102, or 2N4416 transistor
T1, T2	15 turns wound on T37-10 core, three turns secondary
T3	12-V 300-mA filament transformer
L1	Shield bead on transistor lead

FIGURE 9-6. Schematic of the preselector.

The circuits are mounted on copper-clad board which is attached to the tuning capacitor.

for circuit stability. A small piece of PC board was soldered to the main circuit board to act as a shield. The variable capacitor should also have a shield between the two sections. The transistors and most of the other components are supported by soldering to the capacitor lugs and the surface of the copper-clad board. The coils may be fastened to the board with a small amount of silicon rubber, the type used for bathtub caulk and available at most hardware stores. When winding the toroid cores, each time the wire passes through the center of the core, it is counted as one turn.

The input and output leads to the back panel are connected with miniature 50-Ω coaxial cable. The power supply components are mounted near the back of the box and the ac line switch is brought to the front panel while a switch mounted on the back panel switches the preselector in and out of the circuit.

After the circuit has been wired, some simple dc voltage measurements should be made before connecting the unit to a receiver. The supply voltage should be about 18 V. The voltage at the source of Q1 should be about 1 V. The voltage at the gate of Q2 should be about 5V. While monitoring the voltage at the gate of Q2, tune the variable capacitor from maximum capacitance to minimum capacitance. There should be no change in the gate voltage. Any change is an indication of oscillations, which should be corrected with improved shielding and careful lead dress. As in any RF amplifier, only the minimum amount of lead should be used, as excessive lead length will produce unstable circuits.

Assured that there are no oscillations or dc circuit problems, connect the preamplifier to the receiver and antenna. Tune in a signal in th 10-m band and switch in the preamplifier. Turn the tuning control slowly while listening for a rise in background noise. There may be a noticeable rise in the level of the received signal. Tune just one of the trimmer capacitors on the variable capacitor while adjusting the main tuning control until the maximum gain has been achieved. Switching the preselector in and out should produce a rise in signal strength of at least 4 S units. For receivers that have no S-meter, the improvement in signal strength on the 10-m band should be very noticeable. Tune in a strong signal in the 15-m band with the preselector switched out. Switch the preselector in and tune the main tuning control for a maximum signal. The difference in signal strength should be at least 3 S units. Lack of producing 3 S units of improvement indicates poor tracking in the variable capacitor, and it will be necessary to determine which section has the most capacitance. Place a small 2.5- to 11-pF trimmer in parallel with one section of the tuning capacitor. Tune the trimmer capacitor, and the main tuning capacitor for maximum signal on the 15-m band. Move the trimmer to the other section and repeat the tuning. There should be a significant difference in gain between the locations of the capacitor. The side that produced the least gain is the side that has the excess capacitance. It will be necessary to bend the capacitor plates to produce tracking. Carefully pull the outside plates away from the end that is toward full mesh. Adjust the plates only enough to produce the minimum 3 S units of gain. Be sure that the preamplifier produces at least 4 S units of gain at the 10-m band.

9.5 SIMPLE RTTY TERMINAL UNIT

The most important piece of equipment for RTTY operators after their transmitter or receiver is the terminal unit (Figure 9-7). More than anything else, this unit will determine whether a station will have a good "print" in the presence of fading and noise. Although there are excellent suppliers of terminal units for amateur use, most amateurs prefer to build their own. The unit described is constructed from standard components and provides all the features of expensive terminal units.

There have been many discussions of the pros and cons of limiting versus limiterless RTTY operation. To solve the dilemma, this terminal unit can operate in either mode to take advantage of the characteristics of each type of operation. The input signal is either amplified or limited

The terminal unit can be the first piece of equipment of an RTTY station.

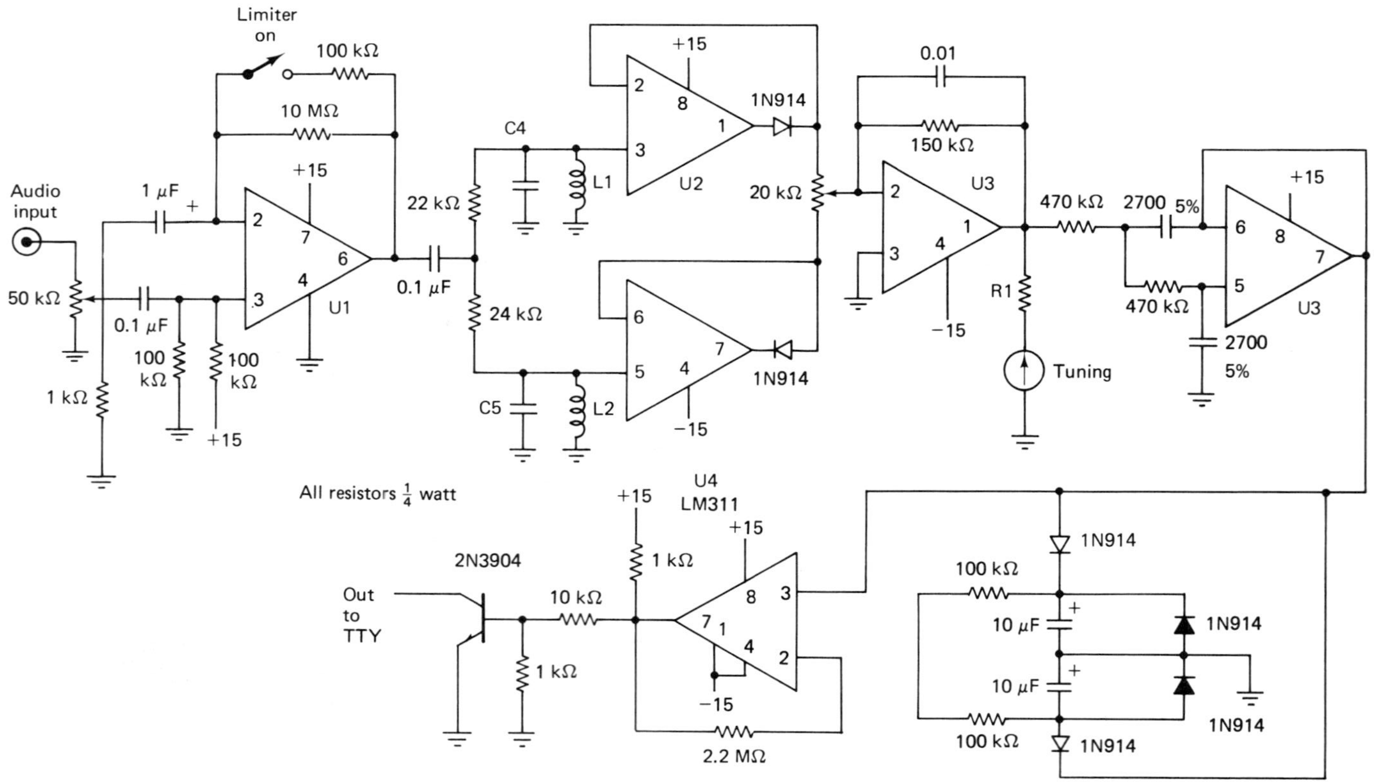

FIGURE 9-7. Schematic of terminal unit.

Parts List for FSK Terminal Unit

C1	1-μF electrolytic, 25 V
C2	0.1-μF ceramic, 50 V
C3	0.1-μF ceramic, 50 V
C4	Tuning capacitor 0.047-μF plus trimmer capacitors
C5	Tuning capacitor 0.047-μF plus trimmer capacitors
C6	0.01-μF ceramic, 50 V
C7	2700-pF mica
C8	2700-pF mica
C9	10-μF electolytic, 25 V
C10	10-μF electtolytic, 25 V
R1	10-kΩ for 1-mA meter
U1	CA3130
U2	MC1458
U3	MC1458
U4	LM311

FIGURE 9-7. Continued.

by a CA3130 operational amplifier which feeds the tuned circuits of the discriminator. One tuned circuit feeds a precision half-wave diode connected as a positive voltage generator, while the other tuned circuit drives a precision rectifier of the opposite polarity. These two voltages are summed across a 20-kΩ potentiometer, which drives a simple low-pass filter. The tuning meter is connected at this point so that it will have a fast response time. The dc voltage is further filtered with another low-pass filter before it is fed to the comparator. When operating with a limiterless demodulator circuit, some method of correcting for selective fading must be incorporated. This terminal unit has an automatic threshold control that remembers the peak of the mark or space tone for a few seconds during the time of signal fading. The value of the peak is adjusted automatically if the amplitude of the signal should change. The automatic threshold is required for limiterless operation but is of no value during limiter operation. Since the threshold correction does not interfere with the limiter type of operation, it is left connected at all times.

This converter uses 88-mH toroid inductors rather than an active filter. The toroids offer a level of performance that is available from a good-quality active filter type of terminal. The simpler inductor circuit was used for ease of tuning and because of sheer simplicity. Toroids of 88 mH have been, and continue to be, an easily obtained item from surplus sources.

Except for the discriminator tuned circuits, there are no difficult adjustments for the terminal unit. To tune the discriminator, apply a

2125-Hz signal from a signal generator to the terminal unit input. The frequency of the test signal must be set to within about 10 Hz, which requires a frequency counter or other method of determination. RTTY textbooks often offer methods of determining the frequency of the standard mark and space tones for use in tuning terminals units if a counter is not available. Using an oscilloscope or a voltmeter, measure the voltage at pin 2 of U2 while tuning the frequency of the audio generator. Set the generator to the peak voltage and determine the resonant frequency. This frequency should be higher than 2195 Hz. Add 0.001- or 0.0047-μF capacitors until the frequency of resonance becomes 2195 plus or minus 20 Hz. Similarly, measure the resonant frequency of the other tuned circuit by monitoring the voltage at pin 7 of U2. This frequency should be tuned to 2125 Hz plus or minus 20 Hz. When tuning the discriminator, the value of the capacitor placed in parallel with the inductor should become progressively smaller, as the resonant frequency converges to the correct value. It is not unusual to have four or more capacitors in parallel with the inductor of the tuned circuit.

The larger the capacitor, the more stable it should be with temperature variations. The largest capacitor should be polystyrene or polypropylene. The capacitors used to trim the circuits can be Mylar or ceramic.

The two toroid inductors dominate the inside of the terminal unit.

Once the toroid tuned circuits are adjusted, the discriminator should be balanced, using R1. Using the same signal generator, tune for a maximum deflection of the front-panel meter for the mark signal. Make a note of this reading and tune the signal generator for the maximum deflection of the meter for the space frequency. If the maximum deflections are not the same, decrease the level of the higher reading by half of the difference by using R1. Check the results by tuning the generator first to mark and then to the space, looking for equal movement of the meter.

Using the terminal unit is very easy and completely automatic. The CRT display or the local loop is connected to the output transistor. When using the terminal unit with a high-voltage local loop, great care must be taken to ensure that the terminal is connected to the ground side of the loop.

Audio may be fed from the speaker output of the receiver, which is usually sufficient to drive the terminal unit into limiting while in the limiterless mode. The only danger with this method of interconnect is that the volume level from the speaker may not be a comfortable listening level, and in addition, changes in the volume will affect the terminal unit. A superior method is to remove audio from the receiver before the volume control so that there is no effect from the volume control setting. This, too, can be a problem because often the voltage level before the volume control is insufficient for reliable terminal operation. If this is the case, an additional amplifier can be added ahead of the terminal unit. Any operational amplifier will provide sufficient gain and bandwidth.

Using the terminal unit is simple. A RTTY signal is tuned in and adjusted so that the meter swings equally from left to right. For most accurate tuning, the meter should be centered while printing a series of RYs. Often, stations calling CQ, starting a tape, or when requested will send a series of RYs for tuning purposes.

When operating in the limiterless mode, the audio gain control must be set low enough to assure that the amplifier is not being overloaded and, in essence, limiting. If advancing the audio control causes the meter to swing farther to the left or to the right, the amplifier is not limiting. If an increase in signal level does not cause an increase in meter deflection, the amplifier is clipping and thus limiting.

Normally, the terminal is used in the limiterless mode. If the level of interference is low and the signal is out of the noise, but contains some fading, the limiter mode will provide a superior print and that mode should be selected.

INDEX

INDEX